Production Technology for Vegetables and Spices

Production Technology for Vegetables and Spices

Avdhesh Kumar Patel
JNKVV-Krishi Vigyan Kendra, Dindori (M.P.)

Radhika Negi
Department of Vegetable Science and Floriculture
CSK HPKV, KVK, Lahaul, and Spiti at Kukumser

Namita Raut
Department of Vegetable Science
University of Horticultural Sciences, Bagalkot, Karnataka

Mohd Ashaq
Department of Botany, Govt Degree College, Thannamandi, J&K

Sheetanshu Gupta
Department of Biochemistry
G.B. Pant University of Agriculture & Technology, Pantnagar, Uttarakhand

Dhirendra Kumar
Department of Botany
Chaudhary Bansi Lal University, Haryana

-EPH-

Elite Publishing House

CRC Press
Taylor & Francis Group
Boca Raton London New York

CRC Press is an imprint of the
Taylor & Francis Group, an **informa** business

First published 2026
by CRC Press
4 Park Square, Milton Park, Abingdon, Oxon, OX14 4RN

and by CRC Press
2385 NW Executive Center Drive, Suite 320, Boca Raton FL 33431

CRC Press is an imprint of Informa UK Limited

British Library Cataloguing-in-Publication Data
A catalogue record for this book is available from the British Library

ISBN: 9781041331223 (hbk)
ISBN: 9781041340621 (pbk)
ISBN: 9781003790938 (ebk)

DOI: 10.1201/9781003790938

Typeset in A Caslon Pro
by Elite Publishing House

-EPH-

Contents

Preface

"Production Technology for Vegetables and Spices" is a comprehensive guide that explores the critical role of vegetables and spices in human nutrition and the national economy. This book is tailored for students, agricultural professionals, and anyone with an interest in horticulture, offering detailed insights into the cultivation practices, management techniques, and economic importance of various vegetables and spices. As global food security and nutritional health continue to be pressing issues, this book aims to provide readers with the essential knowledge and practical skills needed to improve vegetable and spice production sustainably. The significance of vegetables and spices extends beyond their culinary uses. They are vital sources of essential nutrients, vitamins, and minerals, contributing significantly to a balanced diet and overall health. Moreover, the cultivation of these crops plays a substantial role in the agricultural economy, providing livelihoods for millions of farmers and contributing to the GDP. This book begins by highlighting the importance of vegetables and spices in human nutrition and their economic impact, setting the stage for an in-depth exploration of various cultivation techniques.

The text is organized into detailed sections, each focusing on different aspects of vegetable and spice production. The initial chapters introduce the types of vegetable gardening, tracing their origins, current areas of production, and the significance of these crops in different regions. By understanding the historical and geographical contexts, readers can appreciate the evolution and current trends in vegetable and spice farming. Subsequent chapters delve into the practical aspects of cultivation. Detailed descriptions of improved varieties and best practices for growing key vegetables such as potatoes, tomatoes, cauliflower, onions, okra, bottle gourd, and bitter gourd are provided. These sections cover essential cultivation practices, including optimal sowing times, transplanting techniques, planting distances, fertilizer requirements, irrigation methods, and weed management. Understanding these elements is crucial for achieving high yields and maintaining crop health.

The book also addresses the physiological disorders that can affect vegetable crops, guiding on identifying and managing these issues to prevent yield losses. Disease management is another critical focus area, with detailed discussions on common diseases that affect vegetables and effective strategies for controlling them. The importance of proper harvesting and storage techniques is emphasized to ensure that the produce retains its quality and nutritional value until it reaches the consumer. In addition to vegetables, the book provides an extensive overview of spice cultivation. Spices such as ginger, turmeric, coriander, cumin, fennel, black pepper, and cardamom are explored in detail. These chapters cover the unique cultivation requirements of each spice, from sowing to harvesting, and highlight the specific challenges and solutions associated with their production. The economic importance of spices is also discussed, emphasizing their role in global trade and their contribution to the livelihoods of farmers.

Seed production is another vital topic covered in the book. High-quality seeds are the foundation of successful vegetable and spice farming, and the book offers practical advice on seed selection, production techniques, and storage. Ensuring the availability of good-quality seeds can significantly enhance crop yields and sustainability.

Throughout "Production Technology for Vegetables and Spices," readers will find a blend of theoretical knowledge and practical applications. Real-world examples and case studies illustrate successful cultivation practices and innovative techniques being used in different parts of the world. By integrating scientific research with practical farming insights, this book aims to empower readers with the knowledge and skills needed to optimize vegetable and spice production, contributing to food security and sustainable agricultural development.

Authors

About the Authors

Mr. Avdhesh Kumar Patel worked as a Programme Assistant (Horticulture) at JNKVV-Krishi Vigyan Kendra, Dindori (M.P.). He completed his M.Sc. Ag(Horticulture) from JNKVV-Jabalpur (M.P.). He qualified for the NET exam conducted by ASRB. He has more than fifteen years of experience in the field of research and extension. Mr. Patel has received the Young Scientist Award, Young Horticulturist Award, and Excellent Scientist Awards from several societies for his outstanding contribution to the relevant discipline. He has 6 Research Papers, 1 Book Chapter, More than 75 Popular articles, 5 Practical manuals, 1 Radio and 2 TV talks. He has conducted 2 Skill Development Training Programmes of 200 & 210 hours each, 16 farm trials (OFT),16 Front Line Demonstrations (FLDs), and 64 training programs for farmers, rural youth, and extension personnel.

Dr Radhika Negi is presently working as a Subject Matter Specialist (Vegetable Science and Floriculture) at CSK HPKV, KVK, Lahaul, and Spiti 1 at Kukumseri. She has done her post-graduation in Vegetable Science from Dr YSP UHF, Nauni, Solan, and qualified ASRB-ICAR NET in Vegetable Science in 2018. She has 3 years and 8 months of experience in the field of extension and research and is involved in various projects. In recognition of their outstanding contribution to the field of Vegetable Science and Floriculture, she was honored with the Young Scientist Award and begged best paper presentation certificate in 3 days International Conference on "Advances in Agriculture, Veterinary and Allied Science for Improving Livelihood and Environmental Security-2022". She published 7 research papers and review papers/articles in international and national reputed journals, 6 book chapters, 8 popular articles, and 5 abstracts published in different conferences. Besides this, she has conducted and attended numerous training programs (on and off campus), workshops and seminars, etc.

Dr. Namita Raut is presently working as an Assistant Professor of Vegetable Science and Assistant Editor at the Communication Center, University of Horticultural Sciences, Bagalkot, Karnataka. Her previous assignments as Senior Assistant Director of Horticulture served the farming community. Presently working in the University of Horticulture Sciences, Bagalkot. Served for a decade in teaching and guiding PG students. Her research interest is crop improvement of vegetable crops and seed production. Good management Practices in Vegetable Production. Working in Solanaceae and Cucurbitaceae crops and many lines are under developmental process.

Dr. Mohd Ashaq is an Associate Professor of Botany in the Higher Education Department of the Jammu and Kashmir Government presently posted at Govt Degree College Thannamandi, J&K, India. He completed his M.Sc., M.Phil., and Ph. D degrees in Botany with a specialization in Plant Virology from Aligarh Muslim University, Aligarh, India in 1992, 1993, and 1997 respectively. Dr. Ashaq served in Eritrea for 8 years between 2000-2008, finally as Director of Research at Eritrea Institute of Technology, Asmara, State of Eritrea (2006-2008). Dr. Ashaq has 4 books, 8 book chapters, 30 Research papers, 17 patents, and over 50 popular articles to his credit. He has edited over 100 issues of magazines and newsletters for various organizations. Several of his publications on Refugees and volunteerism have been translated into French, Portuguese, and Spanish. He is the commissioned member of 5 of 6 IUCN (International Union of Conservation of Nature) commissions at present. He has presented papers and served as a resource person, invited speaker, and keynote speaker in over 150 conferences/seminars/workshops, etc. He has organized dozens of international and national conferences/seminars, workshops, and FDPs. He is a Reviewer/Editorial board member/ fellow of about three dozen national and international peer-reviewed and Scopus/web of Science journals and societies besides 15 development organizations.

Dr. Sheetanshu Gupta working in Biochemistry, Molecular Biology, and Biotechnology for the last 12 years, an alumnus of G.B. Pant University of Agriculture & Technology, Pantnagar, Uttarakhand, India, ICAR ARS NET (Plant Biochemistry and Plant Physiology) and CSIR NET. Dr. Gupta gained valuable experience in the areas of environmental problems, aeroponics and hydroponics, and protein biochemistry. His work showcased his expertise in applying innovative farming techniques and exploring novel approaches to plant protection. Dr. Gupta researched tissue culture hydroponics and mushroom farming, contributing to the advancement of these fields. Dr. Gupta's academic qualifications, research experience, and expertise in biochemistry and biotechnology make him a valuable asset in the field. With his passion for teaching and research, dedication to innovative farming techniques, and student advocacy, Dr. Gupta continues to contribute significantly to the advancement of agricultural and biochemical sciences with more than 10 published books and nearly 20 Research/ Review papers.

Dr. Dhirendra Kumar is a distinguished Scholar, with more than 15 years of teaching and industrial experience. Currently working as an Assistant Professor in the Department of Botany at Chaudhary Bansi Lal University, Haryana, India. His areas of interest are Plant Biology, Biotechnology, Bioprocess Development, Termite Biology, and Waste Management and currently working on the green synthesis of nanomaterials. In addition, he also received the Outstanding Achievement Award 2021 at the 3rd International Conference (GIAFAS-2021), the Best Young Faculty Award GAREA-2021(Global Academic Research Excellence Award) India, Young Botanist Award 2020" from the Deccan Environmental Research Organization, Bijapur, Karnataka, India. Young Scientist Awards 2015 and 2022 Sponsored by

CGCOST (Chhattisgarh Council of Science and Technology), Chhattisgarh State, GAFEF with Tribhuvan University Nepal and India. Currently, he has completed one University research project, has five granted patents (One Australian and four Indian), Four books, 10 book chapters, and more than 25 research papers in good impact International and National reputed Journals with 20+ national and international Conferences/Seminars paper and oral presentations in India and neighboring countries.

Chapter - 1

Introduction

Since the Green Revolution, India has witnessed substantial advancements in agricultural productivity, with vegetable and spice production being notable highlights. Currently, India is the second-largest producer of vegetables in the world, following China. This milestone underscores India's crucial position in the global food supply chain and emphasizes the significance of vegetables and spices in the Indian diet. The Green Revolution introduced high-yield variety seeds, advanced irrigation techniques, and the increased use of fertilizers and pesticides, significantly boosting crop yields. Vegetables like potatoes, tomatoes, and onions, along with spices such as turmeric, chili, and cumin, have seen significant growth in both production and export. These agricultural achievements not only support the nation's food security but also contribute to the livelihoods of millions of farmers, fostering economic growth and rural development.

Vegetables have long been an integral part of the staple diet in India, a country where approximately 42% of the population adheres to a vegetarian lifestyle. This dietary preference is not only a reflection of cultural and religious practices but also a testament to the recognition of the nutritional value that vegetables offer. They are rich sources of essential nutrients, vitamins, and minerals, making them indispensable for maintaining health and well-being. The growing global trend towards vegetarianism, driven by health considerations, environmental concerns, and ethical reasons, further emphasizes the need to enhance vegetable production to support healthier diets worldwide.

Over the past few years, the consumption of vegetables in India has seen a significant increase, reflecting a broader global trend towards healthier eating habits. Vegetables are highly valued not only for their rapid growth and high yields compared to other crops but also for their critical role in human health. Recognizing their importance, the Indian Council of Medical Research (ICMR) recommends a daily intake of 300 grams of vegetables per person. This recommendation includes 125 grams of leafy vegetables, 100 grams of roots and tubers, and 75 grams of other vegetables. Adequate consumption of vegetables is essential for maintaining overall health and vigor, providing a balanced diet rich in vital nutrients.

DOI: 10.1201/9781003790938-1

Vegetables are defined as herbaceous plants or parts of plants that are typically consumed in their unripe form after cooking. They are categorized based on the part of the plant used, including leaves (such as lettuce, chauli, and palak), stems (like celery), tubers (such as potatoes), and bulbs (like onions and garlic). This classification highlights the diversity of vegetables available, each offering unique flavors, textures, and nutritional benefits.

The significance of vegetables extends beyond their culinary uses. Olericulture, the science of vegetable growing, is a critical branch of horticulture. Historically, vegetables were considered a luxury, but today they are recognized as vital components of the human diet and a significant source of income for growers. Unlike other agronomic crops, vegetables, which are generally short-duration crops, allow farmers to harvest multiple crops from the same area of land within a year. This is particularly advantageous in the context of urbanization, industrialization, and the consequent reduction in agricultural land. Cultivating vegetables can thus be more economical and productive, contributing to food security and economic stability.

Vegetables are often referred to as "protective supplementary foods" due to their rich content of essential amino acids, minerals, vitamins, and nutraceutical substances. These nutrients are vital for the proper functioning of various metabolic processes in the human body. Vegetables are especially rich in minerals like phosphorus, iron, and calcium, as well as vitamins such as vitamin A, vitamin C, and several B vitamins. These nutrients play a crucial role in maintaining good health and vigor.

Vitamins found in vegetables are crucial for numerous bodily functions. Vitamin A, abundant in green leafy vegetables and yellow/orange vegetables like carrots, pumpkins, and sweet potatoes, is essential for vision, immune function, and skin health. Vitamin C, found in vegetables such as bell peppers, broccoli, and tomatoes, is important for collagen synthesis, iron absorption, and immune function. B vitamins, including folate, are vital for energy production, brain function, and DNA synthesis, and are found in leafy greens and legumes.

Minerals are equally important for maintaining health. Iron, present in leafy greens like spinach and legumes, is crucial for the formation of hemoglobin and myoglobin, which transport oxygen in the blood and muscles. Calcium, found in vegetables like kale and broccoli, is necessary for bone health, muscle function, and nerve signaling. Phosphorus, present in a variety of vegetables, is important for the formation of bones and teeth, energy production, and cellular repair.

Dietary fiber, a crucial component found in vegetables, consists of cellulose, non-starch polysaccharides, and lignin. It is essential for maintaining digestive health, lowering cholesterol levels, and regulating blood sugar levels. Fiber also plays a protective role against colon cancer and other gastrointestinal diseases. Vegetables such as beans, lentils, broccoli, and Brussels sprouts are rich in dietary fiber, contributing significantly to a healthy digestive system and overall well-being. Including a variety of high-fiber vegetables in the diet can enhance bowel regularity, prevent constipation, and support weight management by promoting a feeling of fullness.

Antioxidants and phytochemicals in vegetables help protect the body from oxidative stress and inflammation. Compounds like beta-carotene, vitamin C, vitamin E, flavonoids, and polyphenols neutralize free radicals, reducing the risk of chronic diseases such as cancer, heart disease, and neurodegenerative disorders. Vegetables like tomatoes, peppers, spinach, and kale are particularly rich in these beneficial compounds, offering significant health benefits. Regular consumption of antioxidant-rich vegetables can improve skin health, boost immune function, and enhance brain health, making them an essential part of a balanced diet.

Economic Benefits of Vegetable Farming

Vegetable farming provides numerous economic benefits. Due to their short growth cycles and high yield potential, vegetables can offer a steady income for farmers throughout the year. This is particularly crucial in regions with limited land availability, where maximizing productivity is necessary. Vegetables play a vital role in food security, providing a reliable source of nutrition and income for both rural and urban populations. The vegetable sector supports many smallholder farmers, many of whom are women, contributing to rural development and poverty alleviation. Additionally, the vegetable industry generates employment opportunities across various stages of the supply chain, from farming and harvesting to processing, packaging, and distribution. This sector not only bolsters the economy but also ensures the availability of fresh, nutritious produce to a wide population, enhancing overall food security and public health.

Table: High-Fiber Vegetables

S.No	Vegetable	Dietary Fiber Content (per 100g)
1	Beans	6.4g
2	Lentils	7.9g
3	Broccoli	2.6g
4	Brussels sprouts	3.8g

The Role of Spices in Indian Agriculture and Cuisine

India is also renowned globally as the "Land of Spices," producing a vast array of spices that contribute significantly to both domestic and international markets. The country is the world's largest producer, consumer, and exporter of spices. In the fiscal year 2020-2021, India produced approximately 10.1 million tonnes of spices from an area of around 3.9 million hectares. Spices are an integral part of Indian cuisine, imparting flavor, aroma, and color to a wide range of dishes.

The major spices produced in India include pepper, cardamom, ginger, turmeric, cumin, coriander, fenugreek, fennel, mustard, and saffron. These spices are cultivated across different regions of the country, each contributing to the unique flavors and aromas that characterize Indian cuisine and medicinal practices.

Spices are not only important for their culinary uses but also for their medicinal properties. They contain bioactive compounds that have antimicrobial, anti-inflammatory, and antioxidant properties. For example, turmeric contains curcumin, known for its anti-inflammatory and anticancer properties. Ginger is widely used to treat digestive issues and inflammation. Black pepper contains piperine, which enhances the absorption of nutrients and has antioxidant properties.

The spice industry in India is crucial for its economic impact. The domestic market for spices is robust, with significant consumption across households, food processing industries, and traditional medicine. The spice industry also supports millions of farmers and laborers involved in cultivation, harvesting, processing, and marketing.

India is a major exporter of spices, with the total export value of spices reaching USD 3.65 billion in the fiscal year 2020-2021. The major export destinations include the United States, Vietnam, China, Bangladesh, Thailand, the United Kingdom, Germany, and the Netherlands. Spices like chili, turmeric, cumin, cardamom, and pepper are the top exported spices.

The economic benefits of the spice industry are substantial. The sector supports a large number of smallholder farmers and contributes significantly to rural development and poverty alleviation. The spice industry also provides employment opportunities in various stages of the supply chain, from farming and harvesting to processing, packaging, and distribution. Despite the significant progress made in vegetable and spice production, several challenges remain. Post-harvest losses are a major issue, with a substantial portion of production lost due to inadequate handling, storage, and transportation facilities. Effective pest and disease management is also crucial, as crops are susceptible to various pests and diseases that can lead to significant yield losses. Climate change poses a significant threat to production, affecting crop growth and productivity. Efficient water management and maintaining soil health are also critical aspects of sustainable production.

To overcome these challenges and enhance production, several strategies can be implemented. Improving seed quality and developing high-yielding, disease-resistant, and climate-resilient varieties can significantly boost production. Adopting modern farming techniques such as precision agriculture, hydroponics, and protected cultivation can enhance productivity and reduce the impact of adverse environmental conditions. Investing in post-harvest infrastructure, such as cold storage facilities, processing units, and efficient transportation systems, can reduce losses and improve the availability of fresh produce. Value addition through processing and packaging can also increase the shelf life and marketability of produce. Developing efficient supply chains and improving market access for smallholder farmers can enhance their income and ensure a steady supply of vegetables and spices to consumers. Establishing farmer cooperatives, contract farming arrangements, and digital marketplaces can facilitate better market linkages.

Providing training and extension services to farmers on best practices in cultivation, pest and disease management, and post-harvest handling can improve productivity and quality. Extension services can also play a crucial role in disseminating knowledge about new technologies and practices. Government policies and programs that support production, research and development, infrastructure development, and market access are essential to promote the growth of the sector. Subsidies, financial incentives, and favorable trade policies can encourage investment and innovation in farming.

Vegetables and spices play a critical role in achieving food security and improving nutrition. They provide essential nutrients that are crucial for growth, development, and overall health. Increased consumption can help address malnutrition and diet-related diseases, such as obesity, diabetes, and cardiovascular diseases. They are rich sources of micronutrients, including vitamins and minerals, which are often lacking in diets dominated by staple grains. Dietary fiber, abundant in vegetables, is essential for maintaining digestive health and preventing diseases such as colon cancer, heart disease, and diabetes. Antioxidants present in vegetables help protect the body from oxidative stress and inflammation, reducing the risk of chronic diseases such as cancer, heart disease, and neurodegenerative disorders. Promoting the consumption of these foods is essential for improving public health. Nutrition education programs, school feeding programs, and public health campaigns can encourage people to include more vegetables and spices in their diets. Farming can also contribute to sustainable agriculture by promoting biodiversity, improving soil health, and reducing the environmental impact of food production. Practices such as crop rotation, intercropping, and organic farming can enhance sustainability and resilience in production systems.

Chapter - 2

Importance of Vegetables and Spices in Human Nutrition and National Economy

Vegetables and Their Significance in India

India stands as the second-largest producer of vegetables globally, only behind China. The country contributes approximately 10% of the world's total vegetable production. In 2002, India produced 78.2 million tons of vegetables from 5.73 million hectares of land. Indian farmers cultivate an impressive variety of 175 different vegetables, with potatoes, tomatoes, onions, cabbages, and cauliflowers making up 60% of the total production.

Growing Domestic Demand

As India's population grows and dietary habits evolve, the domestic demand for vegetables is expected to rise significantly. Current levels of vegetable production, which range from 83 to 91 million tonnes, are projected to increase to between 151 and 193 million tonnes by 2030. However, Indian farmers are currently unable to meet this high domestic demand, resulting in substantial imports. To bridge this gap, India imports approximately $678 million worth of vegetables annually.

Enhancing Domestic Production

Improving domestic vegetable production is essential to meet future demands. A key area for improvement is the vegetable seed industry. Over 50 seed companies are now focusing on developing new vegetable varieties, particularly high-yielding hybrids. Research and development efforts are led by the Indian Council of Agricultural Research (ICAR), which operates three major institutes dedicated to vegetable research:

1. **Indian Institute of Horticultural Research (IIHR), Bangalore** - Focuses on horticultural crops, including vegetables, fruits, and ornamental plants.
2. **Indian Institute for Vegetable Research (IIVR), Varanasi** - Specializes in the development of new vegetable varieties and technologies.
3. **Indian Agriculture Research Institute (IARI), New Delhi** - Engages in comprehensive agricultural research, including vegetable cultivation.

DOI: 10.1201/9781003790938-2

Nearly all agricultural universities and State Departments of Agriculture are involved in vegetable research and development. Among the 25,000 plant scientists in India, at least 1,000 are dedicated to vegetable research, reflecting the importance of this sector.

Table: Major Vegetable Production Statistics

S.No	Year	Vegetable Production (Million Tons)	Cultivated Area (Million Hectares)
1	2002	78.2	5.73
2	2023	90	6.5
3	2030 (Projected)	151 - 193	7.0 - 7.5

Research and Development in the Vegetable Sector

Research and development in the vegetable sector are vital for enhancing productivity and meeting the increasing demand. The focus on developing high-yielding hybrids and new vegetable varieties is critical. ICAR's efforts, along with the contributions of agricultural universities and State Departments of Agriculture, play a pivotal role in this process. With continuous innovation and development, India aims to reduce its reliance on imports and boost domestic vegetable production to meet future demands.

To encourage the year-round consumption of vegetables, it is essential to tackle the seasonality of production. By processing vegetables, their availability can be extended throughout the year, making them more accessible. However, less than 7% of India's vegetable production undergoes processing. This highlights a significant area for improvement in the vegetable supply chain.

A major issue is the substantial post-harvest losses, with around 20-25% of the vegetables produced being wasted due to inadequate post-harvest handling. For certain vegetables like tomatoes and cabbages, these losses can escalate to as much as 60%. Addressing these challenges requires the development of specialized cold storage facilities, which are becoming more common in metropolitan areas. Additionally, the establishment of dedicated vegetable markets and supermarkets helps in reducing these losses by providing better infrastructure for storage and sales.

Efforts are also being made to set up specialized vegetable marketing centers in key locations. These centers assist farmers in transporting and systematically marketing their produce, thereby minimizing wastage and improving the efficiency of the vegetable supply chain.

Vegetable Exports and Global Trade Opportunities

India is a significant player in the global vegetable market, with an annual export value of approximately $246 million. Historically, Indian vegetable exports were primarily limited to potatoes and fresh onions. However, the Government of India has identified opportunities to expand exports to include various other vegetables such as lettuce, fresh peppers, tomatoes, squash, and gherkins. Notably, the export of fresh tomatoes has seen an eightfold increase, indicating the potential for growth in this sector.

The World Trade Organization (WTO) and the General Agreement on Tariffs and Trade (GATT) have opened new avenues for exporting both fresh and processed vegetables. In the fiscal year 2001-02, India earned Rs. 582.31 crore from the export of fresh vegetables. The Agricultural and Processed Food Products Export Development Authority (APEDA) plays a crucial role in facilitating market access, providing guidance, and promoting agricultural commodity exports. APEDA has identified several traditional vegetables such as okra, bitter gourd, chilli, onion, and potato, as well as non-traditional vegetables like asparagus, celery, sweet pepper, sweet corn, baby corn, green peas, French bean, cucumber, gherkins, and cherry tomato, as having strong export potential.

The cultivation of new vegetable varieties such as gherkin, baby corn, sweet corn, broccoli, Brussels sprouts, Chinese cabbage, asparagus, celery, and parsley is on the rise, offering better returns to farmers. Vegetables being exported include okra, tomato, baby corn, cucumber, gherkins, chillies, French bean, capsicum, bitter gourd, bottle gourd, onion, and potato. Non-traditional vegetables are mainly exported to European countries and Australia, as well as to Gulf and Southeast Asian nations. Whole pod edible garden peas are in demand in European countries, while onions and traditional vegetables are primarily exported to Malaysia, Singapore, Gulf countries, Sri Lanka, Bangladesh, Pakistan, and Nepal.

Processing and Nutritional Importance of Vegetables

Recognizing the global potential of processed fruits and vegetables, the Government of India has designated horticulture development as a priority area. A national policy and various programs have been established to support this sector. Several schemes at both the central and state government levels aim to develop the industry. The Ministry of Food Processing Industries offers significant economic incentives to promote the creation of food parks across the country. Additionally, APEDA and the Ministry of Commerce encourage the establishment of Agri Export Zones (AEZs).

The capacity of India's food and vegetable processing industry has increased from 1.108 million tonnes in 1993 to 2.328 million tonnes as of April 1, 2004. Despite this growth, the utilization of fruits and vegetables for processing in both organized and unorganized sectors remains around 2% of total production. However, there has been positive growth in ready-to-serve beverages, dehydrated and frozen vegetables, tomato products, pickles, spice pastes, and curried vegetables in recent years. Approximately 63 different vegetables are traded globally in dehydrated, canned, and processed forms. Dried and preserved vegetables are in demand in countries such as Egypt, Sri Lanka, the UAE, the USA, and Turkey. In 2003-2004, the export of dried and preserved vegetables from India was estimated to be 211.16 lakh tonnes, valued at Rs. 520.49 crore.

The export of dehydrated vegetables from India is dominated by five main products: onion flakes and powder, tamarind powder, dehydrated vegetables, and garlic powder/flakes. The primary export product is dehydrated onion flakes/powder, with an export volume of about 7,224 MT valued at approximately Rs. 32.95 crores in 2000-01.

Vegetables are an essential source of nutrients necessary for human health, providing carbohydrates, proteins, fats, vitamins, and minerals required for growth and physiological functions. Proteins, carbohydrates, and fats, known as proximate principles, are needed in large quantities and are oxidized in the body to produce energy. Proteins are vital for growth and bodybuilding. Vitamins and minerals, needed in smaller quantities, play crucial roles in physiological processes and metabolic activities Vegetables are rich and comparatively inexpensive sources of vitamins like β-carotene, folic acid, vitamin B, vitamin C, and vitamin E, as well as minerals such as iron, calcium, magnesium, and phosphorus. They also provide a good amount of carbohydrates, proteins (about 4%), and energy (about 10%), making them an integral part of a balanced diet.

Nutritional Significance of Vegetables

Vegetables are an essential part of a healthy diet, offering a variety of nutrients that contribute to overall health and disease prevention. They are rich sources of vitamins, minerals, dietary fiber, and phytochemicals, each playing a crucial role in maintaining health.

1. Vitamins and Minerals

Vegetables provide an array of vitamins and minerals vital for bodily functions and overall health.

Vitamins: Vegetables are excellent sources of numerous vitamins necessary for maintaining bodily functions and overall health.

- **Vitamin A (Beta-Carotene):** *Daucus carota* (carrot) is particularly rich in beta-carotene, a precursor of vitamin A. This vitamin is essential for maintaining good vision, a healthy immune system, and proper functioning of the heart, lungs, and kidneys. Beta-carotene also acts as an antioxidant, protecting cells from damage.
- **Vitamin C:** Found abundantly in vegetables like *Capsicum annuum* (bell peppers) and *Brassica oleracea* (broccoli), vitamin C is crucial for collagen synthesis, which aids in wound healing and maintaining the integrity of skin, blood vessels, and bones. As an antioxidant, it helps protect cells from oxidative stress and supports the immune system.
- **Vitamin K:** Leafy greens such as *Brassica oleracea* (kale) and *Spinacia oleracea* (spinach) are rich in vitamin K, which is essential for blood clotting and bone health. It helps the body make proteins for healthy bones and tissues.
- **B Vitamins:** Vegetables like *Phaseolus vulgaris* (green beans) and *Cucumis sativus*

(cucumbers) contain several B vitamins, including folate (vitamin B9), which is crucial for DNA synthesis and repair, and vital for pregnant women to prevent neural tube defects in infants. B vitamins also play a role in energy production and the formation of red blood cells.

Minerals: Minerals are another critical component of vegetables, necessary for various bodily functions.

» **Potassium:** Found in *Solanum tuberosum* (potatoes) and *Spinacia oleracea* (spinach), potassium is vital for maintaining normal fluid balance, nerve function, and muscle contractions. It helps regulate blood pressure and counteracts the adverse effects of sodium.

» **Magnesium:** Magnesium, present in leafy greens like *Brassica oleracea* (kale), is crucial for over 300 enzymatic reactions, including energy production, protein synthesis, and muscle and nerve function. It also contributes to bone health and the regulation of blood glucose levels.

» **Calcium:** Vegetables such as *Brassica oleracea* (broccoli) and *Spinacia oleracea* (spinach) provide calcium, which is essential for strong bones and teeth, blood clotting, muscle contractions, and nerve signaling.

» **Iron:** Iron is found in dark leafy greens like *Spinacia oleracea* (spinach). It is vital for the formation of hemoglobin, which carries oxygen in the blood, and myoglobin, which provides oxygen to muscles. Iron also supports metabolism and is necessary for growth and development.

2. Dietary Fiber

Dietary fiber is crucial for maintaining digestive health and preventing various chronic diseases.

» **Soluble Fiber:** Soluble fiber dissolves in water to form a gel-like substance, helping to lower blood cholesterol and glucose levels. Vegetables like *Phaseolus vulgaris* (green beans) and *Cucumis sativus* (cucumbers) are rich in soluble fiber. This type of fiber slows digestion, which helps in the absorption of nutrients and promotes satiety, aiding in weight management.

» **Insoluble Fiber:** Insoluble fiber does not dissolve in water and adds bulk to the stool, helping food pass more quickly through the stomach and intestines. It is found in vegetables like *Brassica oleracea* (cabbage) and *Zea mays* (corn). This type of fiber prevents constipation, reduces the risk of diverticular disease, and helps maintain bowel health.

Health Benefits of Dietary Fiber: Dietary fiber has numerous health benefits, including:

» **Improved Digestive Health:** Fiber promotes regular bowel movements and prevents constipation. It also supports the growth of beneficial bacteria in the gut, which can enhance digestion and nutrient absorption.

» **Reduced Risk of Chronic Diseases:** A high-fiber diet is associated with a lower risk of developing chronic diseases such as heart disease, type 2 diabetes, and certain types of cancer. Fiber helps lower cholesterol levels, control blood sugar levels, and reduce blood pressure.

» **Weight Management:** Fiber-rich foods are more filling than low-fiber foods, which can help control appetite and reduce overall calorie intake, aiding in weight management and obesity prevention.

3. Phytochemicals

Phytochemicals are bioactive compounds found in vegetables that offer numerous health benefits.

» **Flavonoids:** Flavonoids are a group of phytochemicals found in a variety of vegetables, including *Allium cepa* (onions) and *Capsicum annuum* (bell peppers). They have antioxidant, anti-inflammatory, and anti-cancer properties. Quercetin, a type of flavonoid found in onions, has been shown to reduce inflammation and lower the risk of chronic diseases.

» **Carotenoids:** Carotenoids are pigments that give vegetables their vibrant colors. They have antioxidant properties and are precursors to vitamin A. *Daucus carota* (carrots) and *Lycopersicon esculentum* (tomatoes) are rich in carotenoids like beta-carotene and lycopene. Lycopene, found in tomatoes, has been linked to a reduced risk of prostate cancer and cardiovascular disease.

» **Glucosinolates:** Glucosinolates are sulfur-containing compounds found in cruciferous vegetables like *Brassica oleracea* (broccoli) and *Brassica napus* (cabbage). They have been shown to have anti-cancer properties by inducing detoxification enzymes, inhibiting tumor growth, and protecting cells from DNA damage.

Health Benefits of Phytochemicals: Phytochemicals provide numerous health benefits, including:

» **Antioxidant Properties:** Many phytochemicals act as antioxidants, neutralizing free radicals that can damage cells and lead to chronic diseases. This helps reduce the risk of conditions like heart disease, cancer, and neurodegenerative diseases.

» **Anti-Inflammatory Effects:** Phytochemicals have anti-inflammatory properties that can help reduce inflammation in the body. Chronic inflammation is linked to various diseases, including heart disease, diabetes, and cancer.

» **Cancer Prevention:** Some phytochemicals have been shown to inhibit the growth of cancer cells and prevent the spread of tumors. For instance, sulforaphane, a compound found in *Brassica oleracea* (broccoli), has been shown to reduce the risk of certain cancers by promoting the elimination of potential carcinogens and inhibiting the growth of cancer cells.

» **Cardiovascular Health:** Phytochemicals like flavonoids and carotenoids support cardiovascular health by reducing blood pressure, improving blood lipid profiles, and enhancing the function of blood vessels.

Vegetables as Protective Foods

The Role of Vegetables in Health and Disease Prevention

Vegetables are vital for maintaining health and preventing numerous diseases. They have a long-standing significance in traditional medicine systems worldwide. Vegetables are packed with therapeutic compounds like antioxidants and flavonoids, which play a crucial role in disease prevention and delaying the aging process. Regular consumption of a variety of vegetables can significantly improve overall health and well-being. Furthermore, these nutrients help in maintaining the body's physiological balance and boosting the immune system, making vegetables indispensable in daily diets.

Antioxidants in Vegetables

Antioxidants are compounds that neutralize free radicals, thereby preventing oxidative stress, which can lead to chronic diseases such as cancer, brain disorders, and arteriosclerosis. Some key antioxidants found in vegetables include:

» β-carotene

» Ascorbic acid (vitamin C)

» α-tocopherol (vitamin E)

» Amino acids

» Flavonoids

These antioxidants are abundant in vegetables like tomatoes, sweet peppers, chillies, and leafy greens. These compounds act as scavengers, protecting cells from damage and promoting overall health. Consuming a diet rich in antioxidant-packed vegetables can enhance skin health, support vision, and improve cardiovascular health by protecting against oxidative damage.

Preventing Specific Diseases with Vegetables

1. **Onion and Garlic:** These vegetables are rich in sulphur compounds such as allicin and diallyl disulphide, which help control blood cholesterol levels and prevent heart diseases. Additionally, they possess strong antibacterial properties, which can help in fighting infections and boosting immunity. Regular consumption of onion and garlic can also improve digestive health and provide anti-inflammatory benefits.
2. **Bitter Gourd:** This vegetable is known for its hypoglycemic ingredient 'charantin,' which is effective in managing diabetes. Consuming bitter gourd can help regulate blood sugar levels, making it a valuable addition to the diet of individuals with

diabetes. Bitter gourd also supports liver health and aids in weight management due to its low-calorie content.

3. **Leguminous Vegetables and Brinjal:** These vegetables are effective in reducing blood cholesterol levels, thus preventing heart diseases. They are rich in dietary fiber, which helps in maintaining healthy digestion and preventing constipation. Including leguminous vegetables and brinjal in the diet can also provide essential vitamins and minerals that support overall health.

4. **Celery:** Contains 3-n-butylphthalide, a compound effective against hypertension. Including celery in the diet can help maintain healthy blood pressure levels and improve heart health. Celery is also a good source of antioxidants and anti-inflammatory compounds, which can help in reducing inflammation and oxidative stress in the body.

5. **Yams:** Certain species of yams contain diosgenin, which is used in the production of cortisone and contraceptive drugs. Diosgenin also has potential anti-inflammatory properties and can support hormonal balance. Consuming yams can provide a good source of energy and essential nutrients, contributing to overall health and well-being.

6. **Cole Crops:** Vegetables like Brussels sprouts, sprouting broccoli, and cabbage contain hydrolyzed glucosinolate products such as isothiocyanates and indoles. These compounds have anti-carcinogenic properties, helping to detoxify carcinogens and providing protection against cancer. Including cole crops in the diet can also improve digestive health and support detoxification processes in the body.

Toxic and Anti-Nutrient Compounds in Vegetables

Despite their health benefits, some vegetables contain toxic and anti-nutrient compounds as natural defense mechanisms. These compounds can persist in cultivated varieties, posing potential health risks if consumed in large quantities or improperly prepared. Understanding and managing these compounds is crucial for maximizing the benefits of vegetable consumption.

Common Toxic Compounds:

1. **Trypsin Inhibitors:** Found in legumes, particularly raw soybeans, these inhibitors affect protein digestibility, potentially leading to nutritional deficiencies. Proper cooking methods can help reduce the levels of trypsin inhibitors, making legumes a safe and nutritious part of the diet.

2. **Phytic Acid:** Present in mature seeds of peas and beans, phytic acid binds to minerals like iron and zinc, reducing their bioavailability and potentially leading to mineral deficiencies. Soaking and fermenting these seeds can help reduce phytic acid content, improving mineral absorption.

3. **Oxalates:** Common in green leafy vegetables and legumes, oxalates can form insoluble

calcium oxalates, leading to the development of urinary stones. Proper preparation methods, such as boiling and consuming in moderation, can help reduce the risk of oxalate-related issues.

4. **Nitrates:** High concentrations of nitrates in leafy vegetables can be converted to nitrites in the gastrointestinal tract, causing methemoglobinemia and the formation of carcinogenic N-nitroso compounds. Ensuring proper agricultural practices and avoiding overuse of nitrogen-based fertilizers can help minimize nitrate levels in vegetables.
5. **Glucosides:** Found in tapioca leaves, glucosides release hydrocyanic acid, which is toxic. Dioscorea species contain dioscorin, a toxic alkaloid that can be neutralized by boiling. Understanding proper cooking techniques is essential to ensure these vegetables are safe for consumption.
6. **Solanin:** Produced by potato tubers exposed to sunlight and some brinjal varieties, solanin is a bitter alkaloid that can be harmful if ingested in large quantities. Storing potatoes in a dark, cool place and avoiding green or sprouted potatoes can help minimize solanin exposure.

Nutritive Value of Vegetables

The following table presents the nutritive value of various vegetables, listing the quantity of key nutrients present in each.

S. No	Nutrient	Vegetable	Quantity Present (per 100 g)
1	Vitamin A (IU)	Bathua leaves	11300 IU
		Colocassia leaves	10278 IU
		Beet leaves	9770 IU
2	Vitamin B1 (Thiamine) (mg)	Chillies	0.55 mg
		Colocassia leaves	0.22 mg
		Tomato (Red)	0.12 mg
3	Vitamin B2 (Riboflavin) (mg)	Fenugreek leaves	0.31 mg
		Amaranthus	0.30 mg
4	Vitamin C (Ascorbic acid) (mg)	Drumstick leaves	220 mg
		Coriander leaves	135 mg
		Chillies	111 mg
		Broccoli	109 mg
		Tomato	31 mg
5	Carbohydrate (%)	Tapioca	38.1 %

		Sweet potato	28.2 %
		Potato	22.6 %
6	Protein (g)	Lima bean	7.9 g
		Peas	7.2 g
		Cowpea	4.3 g
7	Fat (mg)	Potato	118 mg
8	Fiber (g)	Potato	7.52 g
		Chilli	6.8 g
		Amaranthus	1.0 g
		Beet leaf	0.7 g
		Spinach	0.6 g
9	Calcium (mg)	Agathi	1130 mg
		Curry leaves	813 mg
		Amaranthus and fenugreek leaves	395 mg
10	Phosphorus (mg)	Amaranthus	800 mg
		Garlic	187 mg
11	Iron (mg)	Agathi	83.9 mg
		Amaranthus	22.5 mg
12	Potassium (mg)	Spinach	605 mg
		Amaranthus	230 mg

Functions and Deficiency Symptoms of Various Nutrients

The table below summarizes the essential functions and deficiency symptoms of various nutrients, including a comprehensive list of vitamins and minerals. This information highlights the importance of maintaining a balanced diet to ensure adequate intake of these vital nutrients.

S. No	Nutrient	Functions	Deficiency Symptoms
1	Carbohydrates	Supply energy (4 Kcal/g); aid in the assimilation of other nutrients; improve digestion; prevent constipation; reduce cholesterol levels	Loss of energy and weakness
2	Proteins	Formation and maintenance of tissues; body growth; provide energy (4 Kcal/g); form enzymes and hormones for various physiological processes	Retarded growth in children; irritability; apathy; retardation of mental development; skin and hair discoloration; swelling of face, legs, and feet

3	Fats	Supply essential fatty acids; provide concentrated energy (9 Kcal/g)	Lack of energy and weakness
4	Vitamins		
	Vitamin A	Growth support; protection against infections; prevention of night blindness; increases longevity	Night blindness; increased susceptibility to diseases; growth retardation in young individuals
	Thiamine (Vitamin B1)	Maintains appetite; keeps nervous system healthy; helps release energy	Beriberi; loss of appetite; throat enlargement; loss of skin sensitivity; weight loss; body temperature decrease
	Riboflavin (Vitamin B2)	Aids in cell respiration; essential for growth; maintains healthy skin	Cracks at mouth corners; red cracked lips; tongue soreness; oral cavity ulcers; swollen nose; red eyes
	Niacin (Vitamin B3)	Essential for growth; helps release energy	Tongue soreness; pellagra; skin changes on hands, feet, legs, and neck
	Pantothenic Acid (Vitamin B5)	Synthesis and metabolism of proteins, carbohydrates, and fats	Fatigue; depression; irritability; hypoglycemia
	Pyridoxine (Vitamin B6)	Protein metabolism; cognitive development; production of neurotransmitters	Anemia; dermatitis; depression; confusion; weakened immune response
	Biotin (Vitamin B7)	Metabolism of fats, carbohydrates, and proteins	Hair loss; skin rashes; neurological disorders
	Folate (Vitamin B9)	DNA synthesis; cell division; proper fetal development	Megaloblastic anemia; neural tube defects in fetuses; elevated homocysteine levels
	Cobalamin (Vitamin B12)	Red blood cell formation; neurological function; DNA synthesis	Pernicious anemia; neuropathy; memory loss
	Ascorbic acid (Vitamin C)	Necessary for wound healing; iron absorption; formation of cementing substances for cells; strengthens blood vessel walls; essential for bone and teeth calcification	Scurvy; gum and mucous membrane bleeding; tooth decay; increased susceptibility to common colds
	Vitamin D	Builds strong bones and teeth; essential for calcium absorption	Rickets; osteomalacia; dental diseases
	Vitamin E	Antioxidant; helps in normal reproduction	Sterility; neurological problems; muscle weakness
	Vitamin K	Essential for blood clotting; bone metabolism	Prolonged bleeding; bruising; hemorrhagic disease of the newborn

5	Minerals		
	Calcium	Formation of bones and teeth; aids in blood clotting	Rickets; osteomalacia; osteoporosis; muscle spasms
	Iron	Oxygen carrier in the body; formation of hemoglobin	Anemia; pale and smooth tongue; paleness of lips, eyes, and skin; spoon-shaped nails
	Iodine	Formation of thyroxin hormone	Goiter; hypothyroidism; intellectual disabilities in infants
	Magnesium	Enzyme activation; muscle and nerve function; bone health	Muscle cramps; mental disorders; osteoporosis; high blood pressure
	Phosphorus	Formation of bones and teeth; energy production and storage	Weakness; bone pain; fragile bones
	Potassium	Maintains fluid and electrolyte balance; supports cell function	Muscle weakness; cramping; irregular heartbeats; fatigue
	Zinc	Immune function; protein synthesis; wound healing; DNA synthesis	Growth retardation; hair loss; diarrhea; delayed sexual maturation; impotence
	Selenium	Antioxidant protection; thyroid function	Keshan disease (a heart disorder); weakened immune function; infertility
	Copper	Iron metabolism; connective tissue formation; nervous system function	Anemia; bone abnormalities; cardiovascular disease
	Manganese	Bone formation; amino acid, cholesterol, and carbohydrate metabolism	Impaired growth; skeletal abnormalities; impaired glucose tolerance

Spices and Their Significance in India

India's Spice Production and Economic Impact

India is renowned globally as the "Land of Spices," producing a vast array of spices that contribute significantly to both the domestic and international markets. The country is the world's largest producer, consumer, and exporter of spices. In the fiscal year 2020-2021, India produced approximately 10.1 million tonnes of spices from an area of around 3.9 million hectares.

The major spices produced in India include pepper, cardamom, ginger, turmeric, cumin, coriander, fenugreek, fennel, mustard, and saffron. These spices are cultivated across different regions of the country, each contributing to the unique flavors and aromas that characterize Indian cuisine and medicinal practices.

Major Spice-Producing States and Their Contributions

- **Kerala**: Known for black pepper (*Piper nigrum*) and small cardamom (*Elettaria cardamomum*), Kerala contributes significantly to the total spice production in India. The state's hilly terrain and tropical climate are ideal for these crops.
- **Andhra Pradesh and Telangana**: These states are major producers of red chili (*Capsicum annuum*) and turmeric (*Curcuma longa*). The dry and hot climate is conducive to the cultivation of these spices.
- **Tamil Nadu**: Produces a variety of spices including turmeric, black pepper, and coriander (*Coriandrum sativum*).
- **Karnataka**: Known for the production of black pepper, cardamom, and ginger (*Zingiber officinale*).
- **Rajasthan and Gujarat**: Major producers of cumin (*Cuminum cyminum*), coriander, and fennel (*Foeniculum vulgare*).
- **Madhya Pradesh**: Known for large-scale cultivation of garlic (*Allium sativum*) and coriander.
- **Sikkim**: Famous for the cultivation of large cardamom (*Amomum subulatum*).

Domestic and Export Market

India's spice industry is not only crucial for its culinary significance but also for its economic impact. The domestic market for spices is robust, with significant consumption across households, food processing industries, and traditional medicine. The spice industry also supports millions of farmers and laborers involved in cultivation, harvesting, processing, and marketing.

India is a major exporter of spices, with the total export value of spices reaching USD 3.65 billion in the fiscal year 2020-2021. The major export destinations include the United States, Vietnam, China, Bangladesh, Thailand, the United Kingdom, Germany, and the Netherlands. Spices like chili, turmeric, cumin, cardamom, and pepper are the top exported spices.

Export Statistics for Major Spices (2020-2021):

Spice	Quantity (MT)	Value (USD Million)
Chili	498,000	1,032.5
Turmeric	183,000	240.2
Cumin	299,000	547.9
Pepper	28,000	113.8
Cardamom	9,000	134.5
Ginger	75,000	86.4
Coriander	49,000	47.2

Fennel	31,000	41.0
Fenugreek	21,000	30.1
Garlic	30,000	20.5

Source: Spices Board India

Research and Development in Spice Production

The Indian Council of Agricultural Research (ICAR) has several institutes dedicated to the research and development of spices, aiming to improve yield, quality, and resistance to pests and diseases. Notable research institutions include:

» **Indian Institute of Spices Research (IISR)**: Located in Kozhikode, Kerala, it focuses on research related to black pepper, ginger, turmeric, and cardamom.

» **ICAR-National Research Centre on Seed Spices (NRCSS)**: Located in Ajmer, Rajasthan, it focuses on seed spices like cumin, coriander, and fennel.

» **ICAR-Directorate of Medicinal and Aromatic Plants Research (DMAPR)**: Located in Anand, Gujarat, it researches various medicinal and aromatic plants including spices.

These institutions work on developing high-yielding varieties, pest and disease management practices, and sustainable farming techniques. They also collaborate with state agricultural universities and other research bodies to disseminate knowledge and technology to farmers.

Nutritional and Medicinal Importance of Spices

Spices are not only integral to culinary practices but also possess significant nutritional and medicinal properties. They are rich sources of vitamins, minerals, antioxidants, and essential oils that contribute to human health.

» **Black Pepper (*Piper nigrum*)**: Contains piperine, which aids in digestion and has anti-inflammatory properties. It is also rich in vitamins C and K, and minerals like iron and manganese.

» **Turmeric (*Curcuma longa*)**: Contains curcumin, known for its anti-inflammatory, antioxidant, and anticancer properties. It is also used in traditional medicine for treating various ailments.

» **Ginger (*Zingiber officinale*)**: Known for its anti-inflammatory and antioxidant properties. It is used to treat digestive issues, nausea, and inflammation.

» **Cumin (*Cuminum cyminum*)**: Rich in iron, it aids in digestion and boosts the immune system.

» **Cardamom (*Elettaria cardamomum* and *Amomum subulatum*)**: Contains compounds that help in digestion, improve oral health, and have antioxidant and diuretic properties.

- **Coriander (*Coriandrum sativum*)**: Rich in antioxidants, it helps in lowering blood sugar levels and has antimicrobial properties.
- **Fenugreek (*Trigonella foenum-graecum*)**: Known for its antidiabetic, antioxidant, and anti-inflammatory properties. It is also used to improve lactation in nursing mothers.
- **Fennel (*Foeniculum vulgare*)**: Contains antioxidants and antimicrobial properties, and is used to aid digestion and reduce bloating.
- **Garlic (*Allium sativum*)**: Contains allicin, which has antimicrobial, antithrombotic, and cardioprotective properties. It is used to lower blood pressure and cholesterol levels.

Nutritional Significance of Spices

Spices enhance the flavor of food and contribute significantly to human nutrition and health due to their bioactive compounds.

1. **Bioactive Compounds:** Spices are rich in bioactive compounds that offer numerous health benefits. For example, *Curcuma longa* (turmeric) contains curcumin, which has potent anti-inflammatory and antioxidant properties. Similarly, *Zingiber officinale* (ginger) is known for its gingerol content, which aids in digestion and has anti-inflammatory effects.
2. **Antioxidant Properties:** Many spices possess strong antioxidant properties, helping to neutralize free radicals in the body. *Cinnamomum verum* (cinnamon) and *Syzygium aromaticum* (cloves) are notable for their high antioxidant content, which can help protect cells from oxidative stress and reduce the risk of chronic diseases.
3. **Therapeutic Benefits:** Spices have been used in traditional medicine for centuries due to their therapeutic properties. *Capsicum annuum* (chili peppers) contain capsaicin, which is known for its pain-relieving properties and ability to boost metabolism. *Piper nigrum* (black pepper) contains piperine, which enhances nutrient absorption and has anti-inflammatory effects.
4. **Food Security:** Vegetables are essential for ensuring food security. They provide essential nutrients and are a vital part of a balanced diet. Increasing vegetable production can help reduce malnutrition and improve public health, contributing to the overall economic development of a nation.

Spices

The following table presents the nutritive value of various spices, listing the quantity of key nutrients present in each.

S. No	Nutrient	Spice	Quantity Present (per 100 g)
1	Vitamin A (IU)	Paprika	528 IU
		Red chili powder	155 IU
2	Vitamin B1 (Thiamine) (mg)	Coriander seeds	0.239 mg
		Cumin seeds	0.628 mg
3	Vitamin B2 (Riboflavin) (mg)	Cardamom	0.182 mg
		Paprika	0.228 mg
4	Vitamin C (Ascorbic acid) (mg)	Red chili pepper	143.7 mg
		Thyme	160.1 mg
		Parsley	133 mg
5	Carbohydrate (%)	Fennel seeds	52.29 %
		Fenugreek	58.35 %
6	Protein (g)	Mustard seeds	26.08 g
		Coriander seeds	12.37 g
7	Fat (mg)	Sesame seeds	49.67 mg
		Nutmeg	36.31 mg
8	Fiber (g)	Cumin seeds	11.58 g
		Fenugreek	24.6 g
9	Calcium (mg)	Poppy seeds	1438 mg
		Celery seeds	177 mg
10	Phosphorus (mg)	Coriander seeds	370 mg
		Dill seeds	543 mg
11	Iron (mg)	Cloves	11.83 mg
		Turmeric	41.42 mg
12	Potassium (mg)	Basil	295 mg
		Parsley	1000 mg

Economic Importance of Spices

The spice industry significantly contributes to the national economy through employment, export revenue, and the development of value-added products.

1. **Employment and Livelihood:** The spice sector provides livelihoods to millions of

people, especially in spice-producing countries. From cultivation to processing and marketing, the spice industry generates employment opportunities at various levels.

2. **Export Revenue:** Spices are among the most traded commodities globally. Countries like India, known as the land of spices, earn substantial foreign exchange through spice exports. Major spices like *Piper nigrum* (black pepper), *Elettaria cardamomum* (cardamom), and *Cinnamomum verum* (cinnamon) are in high demand internationally.
3. **Value-Added Products:** The spice industry has seen a surge in the development of value-added products such as spice oils, oleoresins, and extracts. These products have applications in the food, pharmaceutical, and cosmetic industries, contributing to economic growth.
4. **Cultural and Culinary Heritage:** Spices hold a significant place in the cultural and culinary heritage of many countries. The global popularity of cuisines rich in spices, such as Indian, Thai, and Middle Eastern cuisines, has boosted the demand for spices, thereby enhancing their economic value.

Vegetables and Spices as General Medicines

Vegetables and spices have long been recognized for their numerous health benefits and curative properties, playing vital roles in traditional and modern medicine. Their rich nutritional content makes them powerful allies in maintaining health and preventing diseases. Here are some notable examples:

» **Obesity Prevention**: Vegetables, being rich sources of proteins, minerals, vitamins, and water while low in fat, help in checking obesity. Spices such as turmeric and ginger also aid in weight management due to their metabolic-boosting properties.

» **Antidote Properties**: Juice from cabbage acts as an antidote against poisonous mushrooms. Similarly, certain spices like clove and cinnamon have detoxifying properties.

» **Anti-Jaundice**: Bitter luffa juice is known as an effective anti-jaundice medicine.

» **Intestinal Disorders**: Carrots have a therapeutic effect on intestinal disorders in children due to the presence of chemicals in the terpene chain in the essential oils found in this plant.

» **Fertility Control**: Extracts from carrots are believed to prevent pregnancy by preventing the implantation of a fertilized egg in the uterus.

» **Antiseptic Properties**: Onions and garlic contain sulfur compounds with antiseptic properties. Garlic, in particular, contains a triterpene compound called garlic inotoldiol, which inhibits fungal growth in vitro conditions.

» **Antipyretic**: Sweet potatoes contain a bitter substance known as ipomeamarone, used by indigenous New Zealanders as an antipyretic.

» **Neurogenic Bladder Treatment:** The pungent extract (capsaicin) of red pepper has been used to treat neurogenic bladder.

» **General Treatments:** The medicinal roles of ginger and turmeric are well documented, especially for their anti-inflammatory and antioxidant properties.

Role of Vegetables and Spices in the National Economy

The horticulture sector, encompassing both vegetables and spices, has demonstrated its immense potential for economic gain, especially in fragile ecosystems. Recognizing its significance, horticulture has been prioritized in the national agricultural scenario to ensure livelihood security and nutritional security for the population.

Vegetables and spices are crucial components of Indian agriculture, contributing to nutritional security due to their short growth cycles, high yields, nutritional richness, and economic viability. They generate significant on-farm and off-farm employment. It is estimated that domestic demand for vegetables and spices will rise substantially. Thus, increasing their production and productivity is paramount for meeting the ever-increasing demand at reasonable prices.

The horticultural landscape in India is rapidly evolving, with a wide range of vegetable and spice crops being cultivated in temperate, tropical, and subtropical regions. Vegetables and spices grow quickly, produce abundantly, and offer high-income potential. However, current production is insufficient to meet the recommended intake of 300 grams of vegetables per person per day, with the current per capita intake being about 145 grams per day. By 2030, the estimated demand for vegetables and spices will range between 151-193 million tons. Given the low daily supply of vegetables per person, it is crucial to increase both production and consumption of vegetables and spices in India by three to four times. The vegetable and spice industries contribute significantly to the national economy, supporting millions of farmers, entrepreneurs, and industries related to seeds, fertilizers, pesticides, herbicides, and machinery.

Vegetables and spices have a substantial impact on income, employment, and nutrition. India's diverse climate allows it to potentially become one of the world's leading exporters of fresh vegetables and spices. These commodities are increasingly seen as vital for food safety and nutrition. Vegetable and spice production offers promising economic opportunities to reduce rural poverty and unemployment, forming an integral part of farm diversification strategies. To fully realize the economic potential of vegetables and spices, governments need to increase their investment in farm production, including improved varieties, alternative pesticides, and protected agriculture. Better harvest management, food security measures, and market access are also crucial.

Promoting the nutritional power of vegetables and spices requires consumer education on their health benefits and ensuring availability at low prices or through self-cultivation.

Consumption should be encouraged through supply-side measures and behavioral change campaigns emphasizing a healthy diet. Governments and donors need to prioritize vegetables and spices, providing increased economic opportunities for farmers and ensuring healthy food for all. This sector holds the potential to significantly uplift emerging farmers and improve food security.

The importance of vegetables and spices in the human diet has been well recognized since time immemorial. They provide essential components of human nutrition and play a crucial role in maintaining health and preventing diseases. As rightly said, "Vegetables and spices are friends of doctors and the glory of cooks." They are vital components of a balanced diet and crucial elements of horticulture, significantly contributing to food and nutritional security. Vegetables and spices serve as effective nutritional supplements, forming a balanced diet for both rich and poor. According to UN food organizations, approximately 25 million people die from hunger and exhaustion every year. In this context, particularly in developing countries, vegetable and spice crops can play an important role in alleviating malnutrition by providing essential minerals, vitamins, and energy in a balanced diet. Vegetable and spice production also offers substantial opportunities for unemployed youth, being highly beneficial for maintaining health and preventing diseases. These crops contain valuable food ingredients that can be utilized to build and repair the body. They play a crucial role in crop diversification, employment generation, and socio-economic upliftment of the farming community.

Vegetables and spices provide carbohydrates, proteins, fats, vitamins, minerals, and water along with roughage, which are essential constituents of a balanced diet. Promoting their consumption and increasing production will significantly contribute to national health and economic stability, ultimately leading to a healthier and more prosperous society.

Chapter - 3

Types of Vegetable Gardening

a. Kitchen Gardening

Kitchen gardening, also known as home gardening or nutritional gardening, involves cultivating vegetable crops within residential areas to fulfill a family's vegetable needs throughout the year. This practice maximizes available space, often in backyards, to grow a diverse range of vegetables for personal consumption.

1. The Role of Vegetables in Nutrition

1.1 Essential Component of Daily Diet

Vegetables play a vital role in daily diets, especially for those adhering to vegetarian lifestyles. They provide essential nutrients, including vitamins, minerals, and dietary fiber, which are crucial for maintaining overall health and well-being. Regular consumption of vegetables is linked to reduced risks of chronic diseases such as heart disease, diabetes, and certain cancers.

1.2 Enhancing Nutritional Value

Incorporating vegetables into meals significantly enhances the nutritive value and palatability of food. Vegetables are rich sources of essential nutrients, helping to meet dietary requirements and improve the taste of meals. For example, leafy greens like spinach and kale are rich in iron and calcium, while carrots and sweet potatoes provide a high amount of beta-carotene, which is converted into vitamin A in the body.

1.3 Dietary Recommendations

Nutrition specialists recommend that an adult should consume 85 grams of fruits and 300 grams of vegetables daily to maintain a balanced diet. However, current vegetable production levels in India only allow for a per capita consumption of 120 grams of vegetables per day, highlighting the need for increased production and consumption. Achieving the recommended intake can help improve overall health and reduce the prevalence of nutrient deficiencies.

DOI: 10.1201/9781003790938-3

2. Benefits of Kitchen Gardening

2.1 Ensuring Self-Sufficiency

Kitchen gardening enables families to grow their own vegetables, ensuring a continuous supply of fresh produce. This practice leverages available freshwater and space, making it a sustainable approach to food production. It also reduces dependency on external sources and ensures that the produce is organic and free from harmful chemicals.

2.2 Effective Pest and Disease Control

Cultivating vegetables in small areas allows for easier control of pests and diseases. Manual removal of affected parts reduces the need for chemical pesticides, ensuring safer produce. This organic approach helps maintain the ecological balance and promotes the use of natural pest control methods.

2.3 Healthier Food Options

Home-grown vegetables are free from harmful pesticide residues, providing a healthier alternative to commercially grown produce. This practice supports organic farming principles and promotes better health by reducing the intake of toxins. Moreover, fresh vegetables retain more nutrients compared to those that have been stored and transported over long distances.

2.4 Cost-Effective

Growing vegetables at home reduces the expenditure on purchasing them from markets. It also contributes to household savings by providing a steady supply of fresh, nutritious produce. The initial investment in seeds, soil, and tools can be offset by the savings from not having to buy vegetables regularly.

2.5 Educational Value

Engaging in kitchen gardening educates family members, especially children, about the importance of agriculture and the value of labor. It fosters an appreciation for the food they consume and encourages healthier eating habits. Children learn about the growth cycle of plants, the importance of sustainable practices, and the benefits of consuming fresh produce.

3. Site Selection for Kitchen Gardens

3.1 Optimal Location

The ideal site for a kitchen garden is typically the backyard of the house. This location is convenient for regular care and maintenance, ensuring that the garden receives adequate attention. A site that receives plenty of sunlight and is close to a water source is preferred to facilitate irrigation and plant growth.

3.2 Efficient Water Utilization

Utilizing wastewater from kitchens and bathrooms for irrigation is an effective way to manage water resources. This practice helps in conserving water while providing necessary hydration for the plants. Implementing rainwater harvesting systems can also support irrigation needs.

3.3 Garden Size

The size of a kitchen garden depends on the availability of land and the number of family members it needs to support. A rectangular plot is preferred over other shapes for better space utilization and ease of maintenance. A well-planned garden layout can maximize the use of available space and ensure a diverse range of vegetables.

3.4 Succession and Intercropping

Implementing succession cropping and intercropping techniques can maximize the use of space and ensure a continuous supply of vegetables throughout the year. With proper planning, a garden of 200 square meters (five cents) can suffice for a family of four to five members. These practices also help in maintaining soil fertility and reducing pest and disease incidence.

4. Preparing the Land

4.1 Initial Preparation

Land preparation involves thorough spade digging to a depth of 30-40 cm. This process helps in loosening the soil and removing stones, bushes, and perennial weeds. Proper soil preparation is crucial for healthy root development and plant growth.

4.2 Enriching the Soil

Incorporating 100 kg of well-decomposed farmyard manure or vermicompost into the soil enhances its fertility. This organic matter improves soil texture and provides essential nutrients for the plants. Regular addition of compost or green manure can maintain soil health and productivity.

4.3 Bed Formation

Forming ridges and furrows at appropriate spacings (45 cm or 60 cm) is essential for optimal plant growth. Alternatively, flat beds can also be used, depending on the type of crops being cultivated. Raised beds can improve drainage and prevent waterlogging, especially in areas with heavy rainfall.

5. Sowing and Planting Techniques

5.1 Direct Sowing

Crops like okra, cluster beans, and cowpea can be directly sown on one side of the ridges. Amaranthus, intended for whole plant pull-out and clipping, can be sown by broadcasting seeds mixed with fine sand. Small onions, mint, and coriander can be planted along the bunds.

5.2 Transplanting Seedlings

Seeds of crops such as tomato, brinjal, and chili should be sown in nursery beds or pots a month in advance. Seedlings are transplanted after 30-45 days, with specific spacing for each crop. Immediate irrigation after planting and regular watering are crucial for healthy growth.

5.3 Perennial Plants

Perennial plants should be placed on one side of the garden, usually at the rear end, to avoid shading and nutrient competition with other vegetable crops. This arrangement ensures that perennial plants do not interfere with the growth of annual crops.

5.4 Utilizing Pathways

The edges of walking paths can be used to grow short-duration greens like coriander, spinach, and fenugreek. This efficient use of space maximizes the garden's productivity and ensures a diverse supply of vegetables.

6. Vegetable Seeds Sowing Calendar

A well-planned sowing calendar ensures a continuous supply of fresh vegetables throughout the year. Below is a detailed vegetable seed sowing calendar:

S. No	Vegetable	Growing Season	Days to Maturity
1	Onion	May-Jun	150-160 days
2	Bottle Gourd	Feb-Mar	55-60 days
3	Carrot	Aug-Oct	75-80 days
4	Tomato	Jun-Aug	110-115 days
5	Melon	Feb-Mar	70-85 days
6	Potato	Oct-Dec	70-120 days
7	Cabbage	Sept-Oct	90-100 days
8	Beans	Feb-Mar	45-50 days
9	Broccoli	Aug-Sept	90-100 days
10	Cauliflower (Late)	Aug-Oct	120-125 days
11	Cauliflower (Mid)	Sept-Oct	120-125 days
12	Apple Gourd	Feb-Mar	70-80 days
13	Cucumber	Feb-Mar	50-70 days
14	Corn	Oct-Nov	60-100 days
15	Lettuce	Sept-Oct	45-55 days
16	Radish	Aug-Jan	40-45 days
17	Pumpkin	Jan-Mar	70-75 days
18	Bitter Gourd	Feb-Mar	55-60 days
19	Okra	Feb-Mar	45-50 days
20	Cauliflower (Early)	Mid-June	120-125 days
21	Capsicum	Nov-Jun	95-100 days
22	Spinach	Sept-Nov	60 days

23	Peas	Sept-Nov	55-60 days
24	Beetroot	Oct-Nov	80-90 days
25	Turnip	Oct-Nov	40-50 days

7. Managing a Perennial Plot

7.1 Selection of Perennial Plants

Include perennial plants such as drumstick, banana, papaya, tapioca, curry leaf, and agathi in your garden. These plants provide a continuous supply of produce and require minimal maintenance.

7.2 Succession Cropping

Ensure that crops are grown continuously throughout the year without any break. Succession cropping involves planting different crops in the same plot one after the other, ensuring a steady supply of vegetables.

7.3 Companion Cropping

Grow long-duration crops alongside short-duration ones to maximize space utilization and productivity. Companion cropping also helps manage pests and diseases, as certain plants can repel pests or attract beneficial insects.

b. Market Gardens

Definition and Purpose

Market gardens are small-scale agricultural operations situated within a 15-20 km radius of cities. Their primary purpose is to supply fresh vegetables to nearby local markets. These gardens are strategically located to minimize transportation costs and ensure the freshest possible produce reaches consumers quickly. Market gardens cater to urban demand, which often includes high-value crops, early varieties, and perishable vegetables that do not withstand long-distance transport.

Cropping Pattern

The cropping pattern in a market garden is highly influenced by the demands of the local market. Farmers often prioritize high-value crops and early-maturing varieties to maximize profits. Crops like amaranth, yard-long beans, and bitter gourd are commonly grown because they are in high demand and fetch good prices. The focus is on producing vegetables that can be harvested and sold quickly to meet market needs.

Economic Considerations

Market gardens face high land and labor costs due to their proximity to urban centers. However, these costs are offset by the high prices that fresh produce commands, the reduced transport costs, and the availability of municipal compost for soil enrichment. Despite these advantages, market gardens are under pressure from urban expansion and rising land costs, leading to their gradual decline in big cities.

c. Truck Gardens

Definition and Purpose

Truck gardens are large-scale agricultural operations usually located in rural areas, designed to supply a few specialized vegetable crops to distant markets. These gardens focus on extensive cultivation of selected crops, leveraging the ideal soil and climatic conditions of rural areas to achieve high yields and maintain quality.

Economic Considerations

Truck gardens benefit from relatively cheaper land and labor compared to urban areas. They often employ mechanical cultivation methods, which increase efficiency and reduce labor costs. High yields result in lower production costs per unit. However, the involvement of middlemen in the marketing process and the costs associated with transporting produce to distant markets can reduce net income. Despite these challenges, truck gardens remain viable due to their scale and efficiency.

Vegetable Gardens for Processing

Definition and Purpose

Vegetable gardens for processing are established to supply processing industries with specific vegetables. Crops like tomatoes, peas, potatoes, sprouting broccoli, spinach, lima beans, gherkins, and onions are cultivated primarily for canning, dehydration, freezing, pickling, and other processing methods. These gardens are typically located near processing factories to ensure a steady and reliable supply of raw materials.

Regional Examples

In India, such gardens are found in states like Punjab, where they supply tomatoes and other vegetables to processing industries. The varieties grown are selected based on their suitability for processing. Often, agreements between factory owners and farmers dictate the supply of seeds, cultivation practices, and delivery schedules, ensuring that the processing industry receives consistent and high-quality produce.

Vegetable Gardens for Seed Production

Annual Seed Requirement

India's annual seed requirement for major vegetables is approximately 51,000 tonnes. Although a significant portion of this demand is met through home-saved seeds by farmers, the advancement of improved varieties and hybrids has transformed seed production into a specialized industry that requires substantial expertise. The increased demand for high-quality seeds has driven the need for advanced techniques and knowledge in seed production to ensure that farmers can access the best possible planting materials for their crops.

Production Standards

Vegetables grown for seed production must adhere to stringent standards to ensure the highest quality. These crops are cultivated under optimal soil, climate, and disease-free conditions. To maintain genetic purity, each crop and variety is grown at specified isolation distances. Regular inspections and rouging, which involves the removal of off-type plants, are essential practices to ensure that only the best seeds are produced. Handling seed crops requires specialized knowledge and skills in curing, threshing, cleaning, packaging, and storage. These processes are critical to maintaining the viability and quality of the seeds until they are planted.

Special Systems of Vegetable Farming

1. Cultivation in Protected Structures

Protected structures such as glasshouses, polyhouses, and rain shelters create controlled environments for growing vegetables. These structures protect crops from adverse weather conditions, pests, and diseases, enabling year-round cultivation and higher yields. They are particularly beneficial in areas with harsh climates or during off-season periods. These environments allow for the precise control of temperature, humidity, and light, which can significantly enhance plant growth and productivity. Additionally, protected cultivation can reduce the need for chemical inputs, leading to safer and more sustainable vegetable production.

2. Hydroponics

Hydroponics is a soil-less method of growing plants using nutrient-rich water solutions. This technique allows for the efficient use of water and nutrients, promoting faster plant growth and enabling the cultivation of vegetables in areas with poor soil quality. Hydroponic systems can be established indoors or in greenhouses, providing a controlled environment that optimizes growth conditions and yields. The ability to control all aspects of the growing environment means that hydroponics can produce consistent and high-quality crops regardless of external weather conditions.

3. Floating Cultivation

Floating cultivation involves growing plants on platforms that float in water bodies. This method is particularly useful in regions prone to flooding or with abundant water resources. Floating gardens utilize space efficiently and provide a sustainable way to cultivate vegetables in aquatic environments. These gardens take advantage of naturally nutrient-rich water, which can reduce the need for additional fertilizers. Floating cultivation can also mitigate the risk of soil erosion and land degradation, making it an environmentally friendly farming practice.

4. River Bed Cultivation

River bed cultivation capitalizes on the fertile soil deposited by rivers. Farmers grow vegetables on dry riverbeds during the off-season when water levels are low. This practice leverages the nutrient-rich soil and moisture retained in the riverbeds to grow crops without additional irrigation, making it a cost-effective and efficient method. The natural fertility of riverbeds can enhance crop yields, and this method can be a valuable strategy for farmers in regions with seasonal water availability.

5. Organic Farming

Organic farming emphasizes the use of natural inputs and sustainable practices to grow vegetables. This method avoids synthetic fertilizers and pesticides, focusing instead on compost, green manure, crop rotation, and biological pest control. Organic farming aims to enhance soil health, biodiversity, and ecological balance, producing vegetables that are free from chemical residues and healthier for consumers. The principles of organic farming also promote environmental sustainability by reducing pollution, conserving water, and improving soil structure and fertility. As consumer demand for organic produce continues to rise, organic farming presents a viable and sustainable option for vegetable production.

Cultivation in Protected Structures

In many developed countries, vegetables are cultivated under protected structures such as greenhouses to combat unfavorable weather conditions like extremely low temperatures, snowfall during winter, and frequent rains during summer. These structures help maintain optimal environmental conditions for crop growth, ensuring high productivity and quality. Utilizing these protected environments, farmers can achieve consistent and high-quality yields throughout the year, regardless of external weather challenges.

Greenhouses

Structure and Function

Greenhouses are structured frames covered with transparent or translucent materials, designed to grow crops under controlled environmental conditions. These structures allow solar radiation to penetrate but trap the long-wave thermal radiation emitted from inside, effectively raising the internal temperature by 10-12°C. This "greenhouse effect" is particularly beneficial for vegetable cultivation in cold climates. Additionally, the enclosed environment increases humidity through evaporation from the soil and plants, and the carbon dioxide released by the plants enhances the rate of photosynthesis. Maintaining an optimum CO2 concentration of 700-900 ppm in a greenhouse atmosphere, and potentially enriching it to 1200 ppm, can significantly boost photosynthesis. CO2 is often supplemented using natural CO2 generators or directed through pipes using liquid CO2, creating a highly efficient growing environment.

Types of Protected Structures

Protected structures are classified based on the cladding material used, with common types including glasshouses and polyhouses.

1. Glasshouses

Structure and Function

Glasshouses are framed structures with glass sheets covering the roof and all four sides. The greenhouse effect inside these structures increases the temperature, which can be further regulated with heating systems during winter and cooling devices during summer. Advanced glasshouses are equipped with computerized microprocessor systems to control temperature, humidity, light, and carbon dioxide levels, creating ideal conditions for plant growth. High-value and highly productive crops such as tomatoes, capsicums, cucumbers, musk melons, watermelons, and spinach are commonly grown in glasshouses. The precision control of environmental factors ensures consistent crop quality and maximizes the potential for high market value, making glasshouses an essential tool for commercial vegetable production.

Crop Management

In glasshouses, trailing cucurbits like cucumbers and muskmelons are pruned to encourage vertical growth and are trained using systems like archways, inclined cordons, or vertical cordon systems. Indeterminate F1 hybrid tomatoes are trained to vertical poles. These plants receive optimal nutrients through drip irrigation and foliar sprays. Although the initial setup cost for glasshouses is high, this investment is offset by prolonged harvesting periods, higher yields, and superior quality produce. The controlled environment also reduces the need for pesticides, enhancing the sustainability and safety of the produce.

2. Polyhouses

Introduction

The development of cost-effective polyhouses has been made possible through advancements in petrochemicals and plastics, providing a viable alternative to expensive glasshouses. Polyhouses are widely used in developed countries such as Japan, Spain, Italy, Holland, Israel, and the UK for commercial vegetable production in adverse climatic conditions. These structures make high-tech farming more accessible and affordable for a broader range of farmers, facilitating year-round production and higher crop quality.

Structure and Usage

Polyhouses are large structures constructed from materials such as aluminum, galvanized iron, or locally available wood, with cladding made from ultraviolet-stabilized low-density polyethylene or transparent plastic film. Polyhouses are classified into low-cost, medium-cost, and high-cost categories based on the climate control devices and materials used, allowing farmers to select the level of investment and technology that best suits their needs and financial capabilities.

Types of Polyhouses

1. **Low-Cost Polyhouse:** These structures lack climate control devices and are made from 700 gauge polyethylene sheets supported by bamboo or other materials. They rely on natural ventilation and shading to manage temperature. These polyhouses are suitable for small-scale farmers aiming to improve crop yields without significant investment in technology.
2. **Medium-Cost Polyhouse:** These polyhouses are equipped with fan and pad cooling systems, shade nets, and micro-sprinklers, and use hot air blowers during winter. They are covered with ultraviolet-stabilized polyethylene of 800 gauge thickness on galvanized iron pipes. Medium-cost polyhouses offer a balance between cost and climate control, making them ideal for mid-sized farms.
3. **High-Cost Polyhouse:** These structures feature fiberglass coverings and full climate control systems. Temperature, humidity, light, day length, and winds are automatically managed using computers. High-tech polyhouses also include automatic fertigation systems, sprinklers, misting systems, and fumigation devices. High-cost polyhouses are designed for commercial operations targeting premium markets where high-quality produce can command higher prices.

S. No	Type	Cost per Square Meter	Features
1	Low-Cost Polyhouse	Rs. 125-500	Basic structure, no climate control devices
2	Medium Cost Polyhouse	Rs. 500-1,000	Fan and pad cooling, shade nets, micro-sprinklers
3	High-Cost Polyhouse	Rs. 1,000 and above	Full climate control, computerized systems, advanced features

S. No	Type	Structure Composition	Climate Control Devices	Typical Crops Cultivated
1	Low-Cost Polyhouse	Bamboo or locally available materials, 700 gauge polythene	Natural ventilation and shading	Cabbage, Cauliflower, Tomato
2	Medium Cost Polyhouse	Galvanized iron pipes, 800 gauge polythene	Fan and pad cooling, shade nets, micro-sprinklers	Tomato, Capsicum, Cucumber
3	High-Cost Polyhouse	Fiberglass, advanced materials	Full climate control with computers, fertigation, etc.	High-value crops like Tomato, Spinach, Musk Melon

Popular Polyhouse Designs

Quonset Type

The Quonset-type polyhouse features a semi-round roof with pipe arches or trusses, covered by UV-stabilized polyethylene sheets. The ideal size for this structure is 30-37.5 meters in length, 9.0 meters in width, with a height of 2.5 meters at the sides and 3.5-4.5 meters at the center. This design is popular due to its simplicity and cost-effectiveness. The curved roof facilitates efficient snow shedding in colder regions and provides ample space for air circulation, which helps in maintaining a stable internal environment conducive to plant growth.

Even-Span Type

Even-span polyhouses have roof slopes of equal width that slant uniformly to either side, typically constructed on level land. The ideal height for these structures is 2.5-2.75 meters at the sides and 3.75-4.25 meters at the center. This design is beneficial for maximizing the use of space and ensuring uniform light distribution across the entire growing area. Even-span polyhouses are particularly effective in regions with moderate climates, providing a balanced environment that supports robust plant growth.

Rain Shelters

Rain shelters are low-cost, naturally ventilated structures designed to protect plants from direct rain. They are especially useful in high rainfall regions such as Assam and Kerala. These shelters are covered with UV-stabilized low-density polyethylene film on the roof, with fully open sides to allow for natural ventilation. Rain shelters are crucial for preventing waterlogging and fungal diseases in crops, thereby ensuring better yield and quality of produce. They offer a simple yet effective solution for extending the growing season in areas with heavy rainfall.

Trench Cultivation

Trench cultivation, also known as underground greenhouse technology, is a method used in high-altitude regions with sub-zero winter temperatures. This technique is a simple and economically viable way to grow vegetables during winter, extending the production season from three to eight months. Farmers in the Ladakh and Leh regions of Himachal Pradesh and Jammu and Kashmir commonly use trench cultivation.

Construction

A trench measuring 10 meters in length, 3.3 meters in width, and 1.25 meters in depth is dug and filled with fertile surface soil up to a height of 30 cm. Stone masonry can be used to prevent wall crumbling. Wooden poles are used to support polyethylene sheets over the trench. This setup utilizes the subsoil temperature, which is higher than the atmospheric temperature, to create a conducive environment for plant growth. Vegetables like beet leaf, spinach, vegetable mustard, fenugreek, Swiss chard, Chinese cabbage, celery, parsley, coriander, and mint are well-suited for trench cultivation.

Tunnels

Tunnels are used to initiate early germination of summer crops like cucurbits. During the rainy season, a plastic tunnel can protect mid-season cucumber varieties raised in nurseries. Tunnels are extensively used in the cold desert of Ladakh for raising vegetable nurseries and obtaining early crops. These structures provide a controlled environment that enhances plant growth when external conditions are unfavorable. Plastic tunnels, made using UV-stabilized corrugated or plain fiber-reinforced plastic sheets supported by metallic wires or frames, help protect crops from rain, wind, and extreme temperatures.

Hot Beds and Cold Frames

Hot Beds

Hot beds are structures used for growing vegetables in winter, providing an early start for crops like tomato, cucumber, okra, pepper, cabbage, muskmelon, and watermelon. They can be made of wood, cement, or brick stone and are prepared above ground. The pit is filled with a mixture of well-rotted farmyard manure, ash, and soil, and artificial heating is employed to maintain warmth. Hot beds create a warm microenvironment that accelerates seed germination and seedling growth, allowing for earlier planting and extended growing seasons.

Cold Frames

Cold frames are similar to hot beds but do not use artificial heating. They are wooden or concrete structures with plastic covers that trap heat from the sun, raising the internal temperature several degrees higher than the outside. Cold frames are useful for extending the growing season and protecting young plants from frost. They provide a cost-effective way to start seedlings earlier in the spring and extend the harvest season into the fall.

Hydroponics

Hydroponics is a soil-less method of growing crops using nutrient solutions. This technique is suitable for areas where soil is lacking or contaminated with disease-causing organisms or toxic substances. Hydroponics is particularly effective for growing high-value vegetables such as tomato, capsicum, lettuce, watermelon, and muskmelon.

Components

- The basic components of a hydroponics system include:
- **Shallow fiberglass trays or plastic trays (60 cm length x 40 cm width x 12.5 cm depth)** for growing plants.
- **A collection tank or nutrient solution storage tank.**
- **A pump** to circulate the nutrient solution from the reservoir to the growing trays through polyethylene tubes.
- **A sequential timer** to control the operation of the pump.

» **An aerator** connected to the reservoir through polyethylene tubes to maintain oxygen levels in the nutrient solution.

Seedlings are planted on trays covered with thermocol sheets, providing support and keeping roots and nutrient solutions in darkness. A uniform slope is maintained in trays to ensure the free flow of nutrient solutions and to avoid water stagnation. Hydroponics systems allow for precise control over the growing environment, leading to faster growth rates, higher yields, and the ability to cultivate crops in non-traditional settings, such as urban areas or regions with poor soil quality.

S. No	Salt Used	Element Provided	Concentration (ppm)
1	Ca(NO3)2	Nitrate N and Calcium (Ca)	NO3 900-930, Ca 180-200
2	KH2PO4	Phosphorus (P) and Potassium (K)	HPO4 90-97, K 234-250
3	MgSO4	Magnesium (Mg) and Sulfate (S)	Mg 49-52, SO4 185-195
4	MnCl2	Manganese (Mn)	Mn 0.02
5	CuSO4	Copper (Cu)	Cu 0.03
6	ZnSO4	Zinc (Zn)	Zn 0.06
7	H3BO3	Boron (B)	B 0.60
8	H3(MO3)2O3	Molybdenum (Mo)	Mo 0.02
9	Ferric citrate	Iron (Fe)	Fe 8.00

Floating Cultivation in Lakes

Introduction

In the Dal lakes of Kashmir, vegetables are cultivated on floating platforms made from the roots of Typha grass, which grows wild in the lakes. This innovative method allows farmers to grow crops on water, taking advantage of the lake's resources. These floating gardens are essential for the local food supply, particularly during the summer months when they provide a substantial portion of the vegetables to Srinagar.

Method

The base for these floating gardens is prepared by weaving the roots of Typha grass into a sturdy mat. This mat is then covered with fertile soil rich in humus to create a nutrient-dense growing medium. Seeds are sown on this base, and all intercultural operations, such as watering and weeding, are performed using boats. One of the unique features of these floating gardens is their mobility; they can be moved to different locations within the lake as needed to optimize growing conditions. This method is particularly effective in areas with limited arable land, providing a sustainable and efficient way to cultivate crops.

River Bed Cultivation

Introduction

River bed cultivation is a unique type of farming practiced in the "diara lands" of Uttar Pradesh and Bihar. This method involves growing cucurbits and other vegetables in the fertile river beds or basins of major rivers such as the Yamuna, Ganga, Gomti, Sarayu, and their tributaries. In North India, this form of vegetable cultivation typically occurs during the winter months from November to February. In Kerala, river beds of Pamba and Manimala are used to grow crops like yard-long beans, bitter gourd, and snake gourd. This method leverages the natural fertility of river beds to produce crops during specific seasons.

Method

River beds are formed by the alluvial and deluvial actions of rivers, which deposit fresh silt and clay during the monsoon months, making the land suitable for vegetable cultivation. The top layer of sand may appear unsuitable, but the subterranean moisture from adjacent river streams supports early crop growth. Cucurbits, with their long tap root systems, are particularly well adapted to these conditions.

The process begins with identifying areas where silt has been deposited and where the water level is about 2.0 meters below the surface. After the South West monsoon ends, typically in late October, trenches or pits are dug at specific intervals. Trenches are spaced 2.0-3.0 meters apart, while pits are dug 4.0 meters apart. Each pit measures about 1.0 meter in diameter and 1.0-1.5 meters in depth. The top layer of sand is removed, and the underlying silt is raised to nearly three-quarters of the pit's height, forming a pillar-like structure surrounded by sand.

Seeds are sown in the silt, and most cucurbits are planted in November-December. The trenches are enriched with farmyard manure or cakes before sowing to enhance soil fertility. Grass stubbles are used to protect the plants from chilling winds and to prevent sand from drifting. This method not only utilizes the natural fertility of river beds but also extends the growing season, allowing farmers to produce vegetables even during off-peak agricultural periods.

Benefits and Challenges

Benefits

1. **Utilization of Natural Resources:** Both floating and river bed cultivation make efficient use of natural resources, such as water and fertile silt, to grow crops.
2. **Extended Growing Seasons:** These methods enable farmers to extend the growing season and produce vegetables during times when traditional farming is not feasible.
3. **Mobility and Flexibility:** Floating gardens can be moved to different locations, optimizing growing conditions and ensuring consistent crop production.

Challenges

1. **Labor-Intensive Practices:** Both methods require significant manual labor for setup, maintenance, and harvesting.
2. **Environmental Dependencies:** Success depends heavily on the natural conditions of rivers and lakes, which can be unpredictable and vary from year to year.
3. **Infrastructure Needs:** Proper equipment, such as boats for floating gardens and tools for digging trenches, is necessary to implement these farming methods effectively.

Organic Farming

Organic farming is an agricultural system that prioritizes the use of natural inputs and processes to maintain and enhance ecological balance. It aims to produce food in a way that is harmonious with nature, ensuring environmental sustainability, soil health, and biodiversity. The principles of organic farming guide farmers in adopting practices that support long-term agricultural productivity and environmental health.

Principles of Organic Farming

1. Seasonal Cultivation:

Cultivating crops in their ideal season is a fundamental principle of organic farming. Seasonal cultivation involves planting crops during their natural growing periods, which maximizes their growth potential and yield. This practice ensures that crops grow in conditions that are most suited to their biological needs, such as temperature, humidity, and light. For instance, leafy vegetables like spinach and lettuce thrive in cooler weather, while crops like tomatoes and peppers prefer warmer conditions.

Adhering to seasonal cultivation reduces the stress on plants, making them less susceptible to pests and diseases. When crops are grown out of season, they often face adverse conditions that can weaken them and make them more vulnerable to infestations and infections. This increased susceptibility necessitates the use of chemical interventions, which organic farming seeks to minimize.

Moreover, seasonal cultivation promotes biodiversity. Different crops are planted in different seasons, which helps maintain a diverse range of plant species in the ecosystem. This diversity is crucial for the health of the soil and the overall farm environment. It also helps in managing pest populations by disrupting their life cycles, reducing the likelihood of severe infestations.

Farmers must make minor adjustments to suit specific localities, considering the unique climate conditions and soil types of their regions. These adjustments might include choosing specific crop varieties that are well-suited to local conditions or modifying planting schedules slightly to account for regional weather patterns. By doing so, farmers can optimize crop performance and sustainability.

2. Avoid Off-Season Cultivation:

Avoiding off-season cultivation is another critical principle in organic farming. Off-season cultivation refers to growing crops outside their natural growing periods. This practice often leads to a higher incidence of pests and diseases due to the unfavorable growing conditions that off-season crops face. These conditions can stress plants, making them more vulnerable to attacks and infections.

When crops are grown out of season, they require more inputs, such as water and nutrients, to survive and produce yields. These additional inputs can strain natural resources and increase the environmental footprint of farming. In organic farming, the goal is to minimize the use of external inputs and rely more on natural processes and cycles.

By sticking to the natural growing seasons, farmers can reduce the need for chemical interventions and enhance the sustainability of their farming practices. This approach helps in maintaining soil health and fertility, as the soil is not subjected to continuous cropping and can recover during off-seasons. It also allows beneficial organisms in the soil to thrive, contributing to the overall health of the farm ecosystem.

Avoiding off-season cultivation also supports the principle of crop rotation, which is essential for preventing the buildup of pests and diseases in the soil. When crops are rotated according to their natural seasons, it disrupts the life cycles of pests and pathogens, reducing their populations and the likelihood of severe outbreaks.

In addition, off-season cultivation can have economic implications. Crops grown out of season often require more labor and resources, increasing production costs. These higher costs can reduce the profitability of farming operations. By adhering to seasonal cultivation, farmers can achieve better economic outcomes while maintaining the ecological integrity of their farms.

3. Variety Selection:

Variety selection is a key principle in organic farming, emphasizing the use of medium-yielding, pest, and disease-tolerant varieties over high-yielding ones. This practice is essential for achieving sustainable agricultural productivity and environmental health. Selecting the right crop varieties involves choosing those that are well-adapted to local conditions and can thrive with minimal external inputs.

Medium-yielding varieties are often more resilient and adaptable than high-yielding varieties. High-yielding crops typically require intensive management and higher inputs of water, fertilizers, and pesticides to achieve their potential. In contrast, medium-yielding varieties are bred for robustness and can perform well under organic farming conditions, which prioritize natural inputs and processes.

Pest and disease tolerance is another critical factor in variety selection. Crops that are naturally resistant to common pests and diseases can significantly reduce the need for chemical interventions. This aligns with the principles of organic farming, which aim to minimize the use of synthetic pesticides and promote a healthy farm ecosystem. For example, certain tomato varieties are resistant to blight, a common fungal disease, reducing the need for fungicidal sprays.

Selecting varieties that are adapted to local conditions is also important. Local varieties have evolved to thrive in specific climates and soil types, making them more resilient to environmental stresses. These varieties often have traits that enhance their survival and productivity in the local context, such as drought tolerance or resistance to local pests.

In organic farming, the focus is on achieving a balance between yield and resilience. While high-yielding varieties might produce more under optimal conditions, they are often less resilient to stress and can suffer significant yield losses in challenging environments. Medium-yielding varieties, on the other hand, offer more stability and consistent performance, contributing to the overall sustainability of the farming system.

Variety selection also plays a role in promoting biodiversity. By growing a diverse range of crop varieties, farmers can enhance the genetic diversity of their farms. This diversity is crucial for building resilience against pests, diseases, and environmental changes. It also supports a healthy farm ecosystem by providing habitats for beneficial organisms and promoting soil health.

4. Resistant Varieties:

Selecting pest and disease-resistant crop varieties is a cornerstone of organic farming. This principle is particularly important in areas prone to severe pest or disease outbreaks, where the use of resistant varieties can significantly reduce the need for chemical interventions. Resistant varieties have natural defenses against specific pests and diseases, making them a crucial component of sustainable farming practices.

The use of resistant varieties aligns with the organic farming goal of minimizing synthetic pesticide use. Chemical pesticides can have detrimental effects on the environment, human health, and beneficial organisms. By planting resistant varieties, farmers can reduce their reliance on these chemicals and promote a healthier farm ecosystem.

Resistant varieties are bred to possess traits that deter or withstand attacks from pests and pathogens. These traits can include physical barriers, such as thick or hairy leaves that make it difficult for insects to feed, or chemical defenses, such as the production of compounds that are toxic or unpalatable to pests. For example, certain squash varieties produce cucurbitacin, a compound that deters cucumber beetles.

In addition to pest resistance, disease-resistant varieties are essential for managing plant health. Diseases like blight, mildew, and rust can devastate crops, leading to significant yield losses. By using varieties that are resistant to these diseases, farmers can ensure healthier plants and more stable yields. This resistance often comes from the incorporation of specific genes that provide immunity or tolerance to pathogens.

The selection of resistant varieties also contributes to the sustainability of the farming system. Healthy, resistant plants are better able to compete with weeds, utilize nutrients efficiently, and withstand environmental stresses such as drought or extreme temperatures. This resilience enhances the overall productivity and stability of the farm.

Moreover, resistant varieties can help in managing pest and disease pressure over the long term. Continuous planting of susceptible varieties can lead to the buildup of pest and disease populations, creating a cycle of increasing infestations and chemical use. Resistant varieties disrupt this cycle by reducing the population of pests and pathogens that can survive and reproduce.

5. Crop Rotation:

Crop rotation is a foundational principle in organic farming that involves growing different types of crops in the same area in sequential seasons. This practice plays a crucial role in maintaining soil health, preventing pest and disease buildup, and enhancing overall farm productivity.

One of the primary benefits of crop rotation is the improvement of soil structure and fertility. Different crops have varying root structures and nutrient requirements. For example, deep-rooted crops like carrots can break up soil compaction and improve soil aeration, while legumes such as beans and peas can fix nitrogen in the soil, enhancing its fertility. By rotating crops, farmers can ensure a more balanced use of soil nutrients and prevent the depletion of specific nutrients.

Crop rotation also helps in breaking pest and disease cycles. Many pests and pathogens are host-specific, meaning they thrive on particular crops. When the same crop is planted repeatedly in the same area, these pests and diseases can build up in the soil, leading to increased infestations and infections. Rotating crops disrupts the life cycles of these pests and pathogens by removing their preferred host, thereby reducing their populations and the need for chemical controls. Additionally, crop rotation can enhance weed management. Different crops have different growth habits and canopy structures, which can influence weed competition. For instance, a dense canopy of leafy vegetables can suppress weed growth by shading the soil, while cover crops like clover can outcompete weeds for nutrients and space. By varying the types of crops grown, farmers can create an environment that is less conducive to weed proliferation.

Crop rotation also supports biodiversity on the farm. Growing a diverse range of crops promotes a healthy ecosystem by providing habitats for beneficial organisms, such as pollinators and natural predators of pests. This biodiversity can enhance ecosystem services, such as pollination and natural pest control, which are essential for sustainable farming.

Implementing crop rotation requires careful planning and knowledge of crop families and their interactions with soil and pests. Farmers need to design rotation schemes that consider factors like nutrient requirements, root depth, and susceptibility to specific pests and diseases. For example, a typical rotation might include a nitrogen-fixing legume crop followed by a heavy-feeding vegetable like corn, then a deep-rooted crop like carrots, and finally a cover crop to restore soil health.

6. Green Manuring:

Green manuring is an essential principle in organic farming that involves growing leguminous crops and plowing them back into the soil before they flower. This practice enriches the soil with organic matter and nutrients, particularly nitrogen, which is crucial for plant growth. Green manuring is a sustainable way to improve soil fertility and structure, supporting healthy crop production without relying on synthetic fertilizers.

Leguminous crops, such as clover, alfalfa, and vetch, are commonly used for green manuring because they have the unique ability to fix atmospheric nitrogen through symbiotic relationships with nitrogen-fixing bacteria (Rhizobium) in their root nodules. This biological nitrogen fixation process converts atmospheric nitrogen into a form that plants can readily use, increasing the nitrogen content of the soil.

Green manuring practice on a farm

When green manure crops are plowed back into the soil, they decompose and release nitrogen and other nutrients, making them available to subsequent crops. This process also adds organic matter to the soil, which improves soil structure, enhances water retention, and promotes the activity of beneficial soil microorganisms. Improved soil structure and increased organic matter content enhance root growth and nutrient uptake, leading to healthier and more productive crops.

Green manuring also helps in weed suppression. The dense growth of green manure crops can outcompete weeds for light, water, and nutrients, reducing weed pressure in the field. Additionally, when these crops are incorporated into the soil, they create a mulch layer that further suppresses weed germination and growth.

Another benefit of green manuring is its role in soil erosion control. The root systems of green manure crops help bind the soil particles together, reducing the risk of soil erosion by wind and water. This is particularly important in areas prone to erosion, where maintaining soil integrity is crucial for sustainable agriculture.

Green manuring also contributes to pest and disease management. Certain green manure crops can produce allelopathic chemicals that inhibit the growth of soil-borne pathogens and pests. Moreover, the practice of incorporating green manure crops into the soil can break pest and disease cycles by disrupting the habitat of specific pests and pathogens.

7. Fallow Periods:

Allowing the land to rest by practicing fallow periods is another essential principle of organic farming. Fallow periods involve leaving a field unplanted for a season or more to restore soil fertility, reduce pest and disease pressure, and improve soil structure. This practice mimics natural ecological processes and helps maintain long-term soil health and productivity.

During fallow periods, the soil undergoes natural processes of regeneration. The absence of crops allows soil microorganisms and beneficial insects to thrive without disturbance. These organisms play a crucial role in breaking down organic matter, cycling nutrients, and improving soil structure. The increased microbial activity during fallow periods enhances soil fertility and health.

Fallow periods also help in breaking pest and disease cycles. Many pests and pathogens are specific to certain crops and can build up in the soil when the same crop is grown repeatedly. By leaving the land fallow, farmers disrupt the life cycles of these pests and pathogens, reducing their populations and the risk of infestations in subsequent crops. This practice is particularly important in organic farming, where the use of chemical pesticides is minimized.

In addition to pest and disease management, fallow periods improve soil structure. Without the constant disturbance of planting and harvesting, the soil can naturally settle and develop better aggregation. This enhanced soil structure improves water infiltration, reduces erosion, and promotes healthy root growth when crops are eventually planted. The natural processes during fallow periods also help in decomposing crop residues, returning nutrients to the soil.

Fallow periods can be managed in various ways to maximize their benefits. Some farmers use cover crops during fallow periods to protect the soil from erosion and add organic matter. Cover crops, such as rye, clover, or buckwheat, are grown and then incorporated into the soil, enhancing its fertility and structure. Others may allow native vegetation to grow, which can provide habitat for beneficial insects and further improve soil health. Practicing fallow periods requires careful planning and consideration of the crop rotation schedule. Farmers need to balance the benefits of fallowing with the need to maintain continuous production. In some cases, partial fallowing, where only a portion of the field is left fallow each season, can be a practical solution that allows for both soil regeneration and ongoing crop production.

8. Composting:

Composting Process

Composting offers numerous benefits to organic farming. Firstly, it recycles waste materials that would otherwise end up in landfills, thus reducing environmental pollution and promoting sustainable waste management practices. By converting waste into a valuable resource, composting aligns with the principles of a circular economy and environmental stewardship. This not only reduces landfill waste but also transforms organic waste into nutrient-rich compost, which can be used to improve soil health and fertility.

Secondly, composting significantly enhances soil structure and fertility. Adding organic matter from compost improves soil aggregation, water retention, and aeration. Improved soil structure supports healthy root growth and nutrient uptake, leading to more vigorous and productive crops. Additionally, the organic matter provides a slow-release source of nutrients, ensuring a steady supply of essential elements for plant growth. Compost also enhances the soil's ability to hold moisture, reducing the need for frequent watering and helping plants withstand drought conditions.

Thirdly, composting promotes soil biodiversity and microbial activity. Introducing compost into the soil increases the population and diversity of beneficial microorganisms such as bacteria, fungi, and earthworms. These organisms play crucial roles in nutrient cycling, organic matter decomposition, and disease suppression. A healthy soil microbiome is essential for sustainable and resilient farming systems. The increased microbial activity also helps to break down organic matter more efficiently, releasing nutrients that are readily available for plant uptake.

Fourthly, composting aids in carbon sequestration. The organic matter in compost adds to the soil's carbon content, sequestering carbon dioxide from the atmosphere and mitigating climate change. This carbon sequestration contributes to the overall environmental benefits of organic farming. By capturing and storing carbon in the soil, composting helps reduce greenhouse gas emissions and combat global warming. Additionally, compost can improve soil health by increasing its organic matter content, which enhances its ability to sequester carbon over the long term.

Organic Inputs

Utilizing locally available organic materials as inputs is a key principle of organic farming. These materials, such as farmyard manure, poultry manure, goat manure, green leaves, organic cakes, fish meal, and bone meal, enrich the soil with essential nutrients and improve its structure. Using organic inputs promotes sustainable farming practices and enhances the health of the farm ecosystem. By relying on locally sourced materials, farmers can reduce their dependence on synthetic fertilizers and support local economies.

Farmyard manure (FYM) is one of the most common organic inputs in organic farming. It consists of decomposed animal dung, urine, and bedding materials, providing a balanced source of nutrients for crops. FYM improves soil fertility, enhances soil structure, and increases microbial activity, supporting healthy plant growth. It also adds organic matter to the soil, improving its water-holding capacity and reducing erosion. Furthermore, FYM helps improve the soil's ability to retain nutrients, making them more available to plants over a longer period.

Poultry manure is another valuable organic input. It is rich in nitrogen, phosphorus, and potassium, essential nutrients for plant growth. Poultry manure is highly effective in boosting soil fertility and promoting vigorous crop growth. However, it should be properly composted or aged before application to avoid the risk of nutrient burn to plants. Properly processed poultry manure can also help reduce the risk of pathogens and weed seeds, ensuring that it is safe and effective for use in organic farming systems.

Goat manure is also beneficial due to its high nutrient content and lower moisture levels compared to other manures. It decomposes quickly and releases nutrients gradually, providing a steady supply of essential elements to crops. Goat manure improves soil texture, increases microbial activity, and enhances nutrient availability. Additionally, its lower moisture content makes it easier to handle and apply, reducing the labor and effort required for its use.

Green leaves and organic cakes, such as neem cake, castor cake, and groundnut cake, are rich sources of organic matter and nutrients. These materials improve soil fertility, suppress soil-borne pests and diseases, and enhance soil health. Organic cakes, in particular, release nutrients slowly, providing a long-lasting source of nutrition for crops. These inputs also have natural pesticidal properties, helping to reduce the need for chemical pesticides and promoting a healthier farm ecosystem.

Fish meal and bone meal are excellent sources of phosphorus and calcium. Fish meal is rich in nitrogen and provides essential amino acids, promoting healthy root and foliage development. Bone meal supplies calcium and phosphorus, supporting strong root growth and overall plant health. Both inputs contribute to balanced nutrient availability in the soil. Additionally, these materials help improve soil structure and microbial activity, further enhancing the soil's ability to support healthy plant growth.

Chapter - 4

Vegetable Classification

A vast array of vegetable crops is cultivated across the country, ranging from large-scale commercial farming to small backyard gardens. These crops exhibit various similarities and differences in their climatic and soil needs, the parts of the plant used, and the cultivation methods. To systematically organize and avoid repetition, vegetable crops can be classified into distinct categories. The different classification methods for vegetables are discussed below.

Botanical Classification

Botanical classification is based on the taxonomical relationships among different vegetables. The plant kingdom is divided into four divisions: Thallophyta, Bryophyta, Pteridophyta, and Spermatophyta. All vegetables fall under the division Angiospermae of Spermatophyta. This division is further subdivided into two classes: Monocotyledoneae and Dicotyledoneae.

Monocotyledoneae includes plants with seeds that have a single cotyledon or seed leaf. These plants typically exhibit parallel venation in their leaves, and their flower parts are usually in multiples of three. Common examples of monocot vegetables include onions (*Allium cepa*), garlic (*Allium sativum*), and asparagus (*Asparagus officinalis*). These plants often thrive in well-drained soils and moderate climates. Monocots usually have fibrous root systems that help stabilize the soil and prevent erosion, making them beneficial for soil conservation practices.

Dicotyledoneae, on the other hand, includes plants with seeds that have two cotyledons. These plants generally show reticulate venation in their leaves, and their flower parts are usually in multiples of four or five. Examples of dicot vegetables include tomatoes (*Solanum lycopersicum*), beans (*Phaseolus vulgaris*), and carrots (*Daucus carota*). These plants can adapt to a wide range of soil types and climatic conditions, making them versatile and widely grown. Dicots typically have taproot systems that can access deeper soil moisture and nutrients, which is particularly advantageous during dry conditions or in nutrient-poor soils. In addition to these broad classes, botanical classification also considers the family, genus, and species of vegetables, providing a detailed understanding of their genetic relationships. For instance, the Solanaceae family includes tomatoes, potatoes (*Solanum tuberosum*), and eggplants

DOI: 10.1201/9781003790938-4

(*Solanum melongena*), which share certain characteristics but also have unique cultivation requirements. Understanding these familial relationships can help farmers and gardeners make informed decisions about crop rotation and pest management, as closely related plants may be susceptible to similar pests and diseases.

Botanical classification not only helps in identifying and categorizing vegetables but also aids in understanding their evolutionary relationships and potential cross-breeding possibilities. This knowledge is crucial for plant breeding programs aimed at developing new vegetable varieties with improved traits such as disease resistance, yield, and nutritional value. For example, breeding programs may focus on creating hybrids that combine the drought tolerance of one species with the pest resistance of another, resulting in more resilient crops. Furthermore, botanical classification can inform conservation strategies for preserving genetic diversity in vegetable crops. By identifying and protecting wild relatives of cultivated vegetables, researchers can ensure that valuable genetic traits are not lost, which could be crucial for future breeding efforts. This genetic diversity serves as a reservoir of traits that can be tapped into to address challenges such as climate change, emerging pests, and diseases.

Overall, botanical classification serves as a foundational framework for the systematic study and cultivation of vegetable crops, ensuring that their diverse characteristics and needs are comprehensively understood and effectively managed. This classification system enables better communication among scientists, farmers, and horticulturists, facilitating the exchange of knowledge and innovations that drive the advancement of sustainable agriculture.

The family-wise distribution of vegetables under these classes is as follows:

Monocotyledoneae

S. No	Family	Botanical Name	Common Name
1	Alliaceae	*Allium cepa*	Onion
2	Alliaceae	*Allium cepa* var. *Aggregatum*	Multiplier onion
3	Alliaceae	*Allium cepa* var. *Viviparum*	Top onion
4	Alliaceae	*Allium porrum*	Leek
5	Alliaceae	*Allium sativum*	Garlic
6	Alliaceae	*Allium fistulosum*	Welsh onion
7	Alliaceae	*Allium ascalonicum*	Shallot
8	Alliaceae	*Allium schoenoprasum*	Chive
9	Liliaceae	*Asparagus officinalis*	Asparagus
10	Araceae	*Dioscorea alata*	Larger yam
11	Araceae	*Dioscorea esculenta*	Lesser yam
12	Araceae	*Colocasia esculenta*	Taro
13	Poaceae	*Zea mays*	Sweet corn

Dicotyledoneae

S. No	Family	Botanical Name	Common Name
1	Aizoaceae	*Tetragonia expansa*	New Zealand spinach
2	Chenopodiaceae	*Beta vulgaris*	Beetroot, Palak
3	Chenopodiaceae	*Beta vulgaris* var. *cicla*	Swiss chard
4	Chenopodiaceae	*Spinacia oleracea*	Spinach
5	Chenopodiaceae	*Artiplex hortensis*	Chakwat, Orach
6	Chenopodiaceae	*Chenopodium album*	Bathua
7	Asteraceae	*Cichorium intybus*	Chicory
8	Asteraceae	*Cichorium endivia*	Endive
9	Asteraceae	*Lactuca sativa*	Lettuce
10	Asteraceae	*Cynara scolimus*	Artichoke
11	Convolvulaceae	*Ipomoea batatas*	Sweet potato
12	Brassicaceae	*Brassica oleracea* var. *acephala*	Kale
13	Brassicaceae	*Brassica oleracea* var. *gemmifera*	Brussels sprouts
14	Brassicaceae	*Brassica oleracea* var. *capitata*	Cabbage
15	Brassicaceae	*Brassica oleracea* var. *botrytis*	Cauliflower
16	Brassicaceae	*Brassica oleracea* var. *italica*	Sprouting broccoli
17	Brassicaceae	*Brassica caulorapa*	Kohlrabi, Knol khol
18	Brassicaceae	*Brassica napus* var. *napobrassica*	Rutabaga
19	Brassicaceae	*Brassica campestris* var. *rapa*	Turnip
20	Brassicaceae	*Brassica juncea*	Leaf mustard
21	Brassicaceae	*Brassica chinensis, B. pekinensis*	Chinese cabbage
22	Brassicaceae	*Armoracia rusticana*	Horse-radish
23	Brassicaceae	*Raphanus sativus*	Radish
24	Cucurbitaceae	*Cucurbita peop*	Summer squash
25	Cucurbitaceae	*Cucurbita moschata*	Pumpkin
26	Cucurbitaceae	*Cucurbita maxima*	Winter squash
27	Cucurbitaceae	*Cucurbita lanatus*	Watermelon
28	Cucurbitaceae	*Cucumis melo*	Musk melon
29	Cucurbitaceae	*Cucumis melo* var. *momordica*	Snap melon
30	Cucurbitaceae	*Cucumis melo* var. *utilissimus*	Long melon
31	Cucurbitaceae	*Cucumis melo* var. *conomon*	Oriental picking melon
32	Cucurbitaceae	*Cucumis sativus*	Cucumber
33	Cucurbitaceae	*Praecitrullus fistulosus*	Tinda

34	Cucurbitaceae	*Sechium edule*	Chow-Chow
35	Cucurbitaceae	*Luffa acutangula*	Ridge gourd
36	Cucurbitaceae	*Luffa cylindrica*	Sponge gourd
37	Cucurbitaceae	*Lagenaria siceraria*	Bottle gourd
38	Cucurbitaceae	*Trichosanthes dioica*	Pointed gourd, Parwal
39	Cucurbitaceae	*Trichosanthes anguina*	Snake gourd
40	Cucurbitaceae	*Momordica charantia*	Bitter gourd
41	Cucurbitaceae	*Benincasa hispida*	Ash gourd
42	Euphorbiaceae	*Manihot esculenta*	Tapioca
43	Fabaceae	*Pisum sativum*	Peas
44	Fabaceae	*Phaseolus vulgaris*	French bean
45	Fabaceae	*Phaseolus lunatus*	Lima bean
46	Fabaceae	*Vicia faba*	Broad bean
47	Fabaceae	*Vigna unguiculata*	Cowpea
48	Fabaceae	*Cyamopsis tetragonoloba*	Cluster bean
49	Fabaceae	*Vigna unguiculata* var. *sesquipedalis*	Asparagus bean
50	Fabaceae	*Lablab purpureas*	Lablab bean
51	Fabaceae	*Glycine max*	Soybean
52	Fabaceae	*Psophocarpus tetragonolobus*	Winged bean
53	Fabaceae	*Tigonella foenum graecum*	Methi, Fenugreek
54	Fabaceae	*Tigonella corniculata*	Kasuri methi
55	Malvaceae	*Abelmoschus esculentus*	Okra, Bhindi
56	Solanaceae	*Solanum tuberosum*	Potato
57	Solanaceae	*Solanum melongena*	Brinjal
58	Solanaceae	*Solanum lycopersicum*	Tomato
59	Solanaceae	*Capsicum annuum*	Chilli
60	Apiaceae	*Daucus carota*	Carrot
61	Apiaceae	*Petroselinum crispum*	Parsley
62	Apiaceae	*Apium graveolens*	Celery
63	Apiaceae	*Pastinaca sativa*	Parsnip

Classification Based on Hardiness

This classification is based on the ability of vegetables to withstand frost and low temperatures, helping determine their appropriate growing seasons. Vegetables are categorized into hardy, semi-hardy, and tender.

S. No	Type	Vegetables
1	Hardy	Asparagus, Crucifers, Garlic, Leek, Onion, Parsley, Peas, Radish, Rhubarb, Spinach
2	Semi-Hardy	Carrot, Celery, Beetroot, Globe artichoke, Lettuce, Palak, Parsnip, Potato
3	Tender	Amaranth, Okra, Brinjal, Chilli, Cluster bean, Cucurbits, Tomato, Colocasia, Amorphophallus, Yams, Sweet potato

Hardy vegetables tolerate frost and low temperatures and are typically winter or cool-season crops. Semi-hardy vegetables can withstand moderate cold but not frost, while tender vegetables are warm-season crops that cannot tolerate frost.

Classification Based on Parts Used

Vegetable crops can also be classified based on the parts used for consumption. This classification helps in understanding the cultural requirements and uses of different vegetables.

S. No	Part Used	Vegetables
1	Tender stem and leaves	Cabbage, Chinese cabbage, Knol khol, Amaranth, Palak
2	Fruits	Tomato, Brinjal, Chilli, Cowpea
3	Flower parts	Sprouting broccoli
4	Underground portion	Carrot, Radish, Beetroot, Potato

Classification Based on Cultural Requirements

This classification groups vegetables with similar cultural requirements together, providing a practical approach for farmers and gardeners.

S. No	Group	Characteristics and Vegetables
1	Solanaceous fruit vegetables	Includes crops like Tomato, Brinjal, and Chilli which are seed propagated, direct sown, and require similar care.
2	Cucurbits	Includes gourd-like vegetables like Bottle gourd, Bitter gourd, and Cucumbers, known for their trailing growth habit and similar cultural practices.
3	Peas and beans	Encompasses Leguminous crops like Peas, French beans, and Cowpeas that fix nitrogen and require similar growing conditions.
4	Cole crops	Covers vegetables like Cabbage, Cauliflower, and Broccoli, which are cool-season crops requiring similar cultivation techniques.

5	Bulb crops	Includes Onion and Garlic, known for their bulb formation and similar cultural practices.
6	Root crops	Consists of Carrot, Radish, and Beetroot, which are grown for their edible roots.
7	Potato	Specifically includes Potato, which requires distinct cultural practices and soil conditions.
8	Tuber crops	Encompasses crops like Sweet potato and Yams, grown for their tuberous roots.
9	Okra	Specifically includes Okra, known for its distinct growth habit and cultural requirements.
10	Pot herbs/greens	Covers leafy vegetables like Spinach and Amaranth, grown for their tender leaves.
11	Salad crops	Includes Lettuce and Endive, typically consumed raw in salads.
12	Perennial vegetables	Covers long-lasting crops like Asparagus and Rhubarb, which remain productive for several years.

Classification Based on Season of Cultivation

This classification groups vegetables based on their optimal growing seasons, helping farmers plan their planting schedules effectively.

S. No	Season	Vegetables
1	Winter season	Cruciferous vegetables (Cabbage, Cauliflower, Broccoli), Carrot, Radish, Beetroot, Onion, Garlic, Peas
2	Mild winter	Hyacinth bean, Winged bean, Tomato
3	Summer season	All gourds (Bottle gourd, Bitter gourd, Ridge gourd), Amaranth
4	Rainy season	Bhindi (Okra), Chilli, Brinjal (Eggplant), Cluster bean, Cowpea

Vegetable crops can also be classified based on other factors such as duration of crop growth (annual, biennial, perennial), ability to grow and set seeds under particular climates (temperate, tropical, sub-tropical), mode of pollination (self-pollinated, cross-pollinated, often-pollinated), and pH requirements of the soil.

These classifications are not rigid, and some crops may fall into multiple categories or require specific adjustments based on local conditions. For instance, while Brinjal and Chilli are typically rainy-season crops, they can also be grown in other seasons with adequate care and suitable varieties. Similarly, crops like Cowpea have varieties adapted to different seasons, emphasizing the need for a flexible approach to classification.

Classification of Spice Crops

Spices are an essential component of culinary arts and play a significant role in enhancing the flavor and aroma of food. Like vegetables, spice crops can be classified in various ways to better understand their cultivation, usage, and characteristics. Below are some common classifications of spice crops:

Botanical Classification

Spice crops can be classified based on their botanical families, which helps in understanding their genetic relationships and cultural requirements.

Monocotyledoneae

S. No	Family	Botanical Name	Common Name
1	Zingiberaceae	*Zingiber officinale*	Ginger
2	Zingiberaceae	*Curcuma longa*	Turmeric
3	Zingiberaceae	*Elettaria cardamomum*	Cardamom
4	Zingiberaceae	*Alpinia galanga*	Galangal
5	Zingiberaceae	*Curcuma zedoaria*	Zedoary

Dicotyledoneae

S. No	Family	Botanical Name	Common Name
1	Piperaceae	*Piper nigrum*	Black pepper
2	Myrtaceae	*Syzygium aromaticum*	Clove
3	Apiaceae	*Cuminum cyminum*	Cumin
4	Apiaceae	*Coriandrum sativum*	Coriander
5	Apiaceae	*Foeniculum vulgare*	Fennel
6	Apiaceae	*Anethum graveolens*	Dill
7	Solanaceae	*Capsicum annuum*	Chili pepper
8	Solanaceae	*Capsicum frutescens*	Cayenne pepper
9	Lauraceae	*Cinnamomum verum*	Cinnamon
10	Lauraceae	*Cinnamomum cassia*	Cassia
11	Lamiaceae	*Ocimum basilicum*	Basil
12	Lamiaceae	*Origanum vulgare*	Oregano
13	Lamiaceae	*Mentha spicata*	Spearmint
14	Lamiaceae	*Mentha piperita*	Peppermint
15	Brassicaceae	*Brassica nigra*	Black mustard
16	Brassicaceae	*Sinapis alba*	White mustard

17	Rutaceae	*Citrus hystrix*	Kaffir lime
18	Fabaceae	*Trigonella foenum-graecum*	Fenugreek

Classification Based on Plant Parts Used

Spice crops can also be classified based on the part of the plant that is used as a spice. This classification is useful for understanding how different parts of plants contribute to flavor and aroma.

S. No	Part Used	Spices
1	Root and Rhizomes	Ginger (*Zingiber officinale*), Turmeric (*Curcuma longa*), Galangal (*Alpinia galanga*)
2	Seeds	Cumin (*Cuminum cyminum*), Coriander (*Coriandrum sativum*), Fennel (*Foeniculum vulgare*), Dill (*Anethum graveolens*), Black mustard (*Brassica nigra*), White mustard (*Sinapis alba*)
3	Bark	Cinnamon (*Cinnamomum verum*), Cassia (*Cinnamomum cassia*)
4	Flowers and Flower Buds	Clove (*Syzygium aromaticum*), Saffron (*Crocus sativus*)
5	Fruits	Black pepper (*Piper nigrum*), Chili pepper (*Capsicum annuum*), Cardamom (*Elettaria cardamomum*), Allspice (*Pimenta dioica*)
6	Leaves	Basil (*Ocimum basilicum*), Oregano (*Origanum vulgare*), Bay leaves (*Laurus nobilis*), Kaffir lime (*Citrus hystrix*)
7	Whole Plant	Mint (*Mentha spicata, Mentha piperita*)

Classification Based on Usage

Spice crops can also be classified based on their usage in culinary practices, medicinal purposes, or other applications.

S. No	Usage	Spices
1	Culinary Spices	Black pepper (*Piper nigrum*), Cinnamon (*Cinnamomum verum*), Turmeric (*Curcuma longa*), Cardamom (*Elettaria cardamomum*), Cumin (*Cuminum cyminum*)
2	Medicinal Spices	Ginger (*Zingiber officinale*), Turmeric (*Curcuma longa*), Clove (*Syzygium aromaticum*), Basil (*Ocimum basilicum*), Mint (*Mentha spicata, Mentha piperita*)
3	Aromatic Spices	Basil (*Ocimum basilicum*), Oregano (*Origanum vulgare*), Spearmint (*Mentha spicata*), Peppermint (*Mentha piperita*), Bay leaves (*Laurus nobilis*)

4	Industrial Spices	Vanilla (*Vanilla planifolia*), Saffron (*Crocus sativus*), Cassia (*Cinnamomum cassia*), Nutmeg (*Myristica fragrans*)

Classification Based on Climatic Requirements

Spice crops can be classified based on their climatic requirements, which helps in planning their cultivation in different regions.

S. No	Climate Type	Spices
1	Tropical Spices	Black pepper (*Piper nigrum*), Cardamom (*Elettaria cardamomum*), Clove (*Syzygium aromaticum*), Ginger (*Zingiber officinale*), Turmeric (*Curcuma longa*)
2	Subtropical Spices	Coriander (*Coriandrum sativum*), Fennel (*Foeniculum vulgare*), Dill (*Anethum graveolens*), Basil (*Ocimum basilicum*), Mint (*Mentha spicata, Mentha piperita*)
3	Temperate Spices	Saffron (*Crocus sativus*), Horseradish (*Armoracia rusticana*), Mustard (*Brassica spp.*)

Chapter - 5

Vegetables

1. Potato (*Solanum tuberosum L.*)

Family: Solanaceae

Origin: Peru and Bolivia in South America

History

The potato likely originated in the central Andean region of South America, specifically in Peru and Bolivia. Historical evidence indicates that South American Indians cultivated potatoes for centuries, using the tubers as a staple food. The Spaniards encountered this valuable plant during their conquests and subsequently introduced it to Europe. Potatoes were transported from Peru to Spain in 1565. It is believed that Sir Francis Drake introduced them to England around 1586.

After its introduction to Europe, the potato quickly became a crucial food crop in various countries, including Italy, France, and Ireland. Its importance was especially notable during the famine years in Ireland, where it provided a vital food source. The potato's adaptability and nutritional value made it a staple in European diets. In the early seventeenth century, the potato was introduced to India from Europe, most likely by the Portuguese, who were among the first to establish trade routes to the East.

The first recorded mention of potatoes in India appears in Terry's account of a banquet at Ajmer hosted by Asaph Khan for Sir Thomas Roe in 1615. By 1675, potatoes were being cultivated in numerous gardens in Surat and Karnataka. Potato farming commenced in the Nilgiri hills in 1822 and subsequently expanded to the northern hills, where it became an important cash crop. The introduction and spread of potato cultivation in India significantly contributed to the agricultural landscape, providing a reliable food source and economic benefit to the farmers.

The potato's journey from South America to Europe and then to India highlights its global significance and adaptability to various climates and soils. In each region, the

DOI: 10.1201/9781003790938-5

potato has played a critical role in sustaining populations, especially during times of food scarcity. Its high yield and nutritional value made it an essential crop in many countries' agricultural systems. Moreover, the potato's introduction to different parts of the world led to the development of numerous varieties adapted to local conditions. This diversification has allowed for the continuous improvement of potato crops, enhancing their resistance to pests and diseases and improving yield and quality. The historical spread of the potato also facilitated cultural exchanges and influenced culinary traditions in the regions it reached.

In modern times, the potato remains a vital crop globally. Advances in agricultural practices and technology have further enhanced its cultivation and productivity. Research continues to focus on improving potato varieties to withstand climate change, ensuring food security for future generations. The potato's enduring presence in global agriculture underscores its importance as a versatile and resilient crop.

Nutritive Value and Uses

Potatoes are a crucial crop globally, serving as a dietary staple in many cultures. In India, they are primarily consumed as a vegetable, often incorporated into mixed vegetable dishes. Additionally, potatoes are used in a variety of culinary preparations including chips, halwa, gulab jamun, rasgulla, murabba, kheer, gujia, and barfi, showcasing their versatility in Indian cuisine.

The nutritional composition of potatoes can vary slightly based on the source. According to Aykroyd (1941), the composition includes:

- 74.7% water
- 22.9% sugar
- 0.6% protein
- 0.1% fat
- 0.6% mineral matters
- 0.01% calcium
- 0.03% phosphorus
- 0.0007% iron

In comparison, MacGillivray (1953) provides the following nutritional details:

- 77.8% water
- Energy: 85 calories per 100g
- Protein: 2g
- Calcium: 13mg
- Vitamin A: 40 IU

- » Ascorbic acid: 12mg
- » Thiamine: 0.11mg
- » Riboflavin: 0.06mg
- » Niacin: 1.18mg

Chatfield (1949) offers additional insights into the mineral content of 100g of raw potato:

- » Sodium (Na): 3mg
- » Potassium (K): 410mg
- » Calcium (Ca): 14mg
- » Magnesium (Mg): 27mg
- » Iron (Fe): 0.8mg
- » Copper (Cu): 0.16mg
- » Phosphorus (P): 56mg
- » Sulfur (S): 29mg
- » Chlorine (Cl): 35mg

Potatoes, when exposed to light, can produce solanine, an alkaloid that is toxic. However, peeling potatoes can remove about 70% of solanine, reducing the risk of poisoning. Consuming solanine can cause illness or even death in humans and livestock, emphasizing the importance of proper storage and handling.

Beyond their nutritional benefits, potatoes also have medicinal uses as documented by Nadkarni (1927). They possess anti-scorbutic properties, making them effective against scurvy. Potatoes are also easily digestible, making them suitable for individuals with neurotic and liver dyspepsia. They act as a diuretic and galactagogue, aiding in increased urine production and milk flow in nursing mothers. Additionally, potatoes have a nervous sedative effect, can stimulate gout treatment, serve as an antispasmodic in chronic coughs, and can be used as a plaster for burns to promote healing.

Classification, Botany, and Floral Biology of Potatoes

Classification

The domestication of the potato began in the high plateau regions of the Peruvian and Bolivian Andes, where it has been cultivated for several millennia. Seven cultivated species of potatoes were domesticated in this region, and they remain a vital crop in the high Andean areas, which boast the greatest genetic diversity of cultivated potatoes (Engel, 1970). This genetic diversity is crucial for breeding programs aimed at improving potato varieties to enhance disease resistance, yield, and adaptability to various climates.

The potato plant belongs to the Solanaceae family and is scientifically known as *Solanum tuberosum*. This species is characterized by its underground tubers, which are the edible part of the plant. Potatoes are classified botanically based on their morphological characteristics, which include the structure of the leaves, flowers, and tubers. The botanical classification helps in understanding the evolutionary relationships among different potato varieties and aids in the identification and conservation of genetic resources.

Potatoes exhibit a wide range of morphological diversity, which has been harnessed through selective breeding to develop varieties suited to different environmental conditions and culinary preferences. The diversity within the species allows for the cultivation of potatoes in various climates, from tropical to temperate regions, making them a globally important food crop.

The high genetic diversity of potatoes in the Andean region is not only significant for agricultural purposes but also for cultural reasons. Indigenous communities in these areas have cultivated and maintained a wide variety of potato species, each with unique flavors, colors, and nutritional profiles. This diversity is a testament to the rich agricultural heritage and traditional knowledge of these communities.

The cultivated potato species are as follows:

Species	**Distribution**	**Origin**
Solanum tuberosum subsp. *tuberosum* (2n = 48)	Cosmopolitan	By artificial selection in Europe, North America, and Chile from introduced clones of group 2.
S. tuberosum subsp. *andigena* (2n = 48)	Venezuela to northern Argentina; also sporadically in Central America and Mexico	From group 4 and 5 by spontaneous doubling of the chromosome number.
S. chaucha (2n = 36)	Central Peru to northern Bolivia	By hybridization between group 2 and 4 and 5. By selection for short tuber dormancy from group 5.
S. phureja (2n = 24)	Venezuela to northern Bolivia	By hybridization between group 2 and 4 and 5. By selection for short tuber dormancy from group 5.
S. stenotomum (2n = 24)	Southern Peru to northern Bolivia	By natural hybridization between wild species followed by artificial selection.
S. juzepczukii (2n = 36)	Central Peru to southern Bolivia	From crosses of *S. acaule* with groups of 4 and 5.
S. curtilobum	Central Peru to southern Bolivia	Crosses of *S. juzepczukii* with group 2.

Botany and Floral Biology

Potatoes belong to the Solanaceae family, which includes other economically significant crops such as tomatoes, eggplants, and peppers. The botanical structure of the potato plant includes stems, leaves, flowers, and tubers. Each part plays a crucial role in the plant's growth and development.

Plant Structure

1. **Stems**: The potato plant has both aerial and underground stems. The aerial stems are green and above ground, while the underground stems, known as stolons, give rise to tubers.
2. **Leaves**: The leaves are compound, with several leaflets arranged around a central rachis. They are typically dark green and covered with trichomes.
3. **Flowers**: Potato flowers are typically white, pink, or purple, and are borne in clusters. The flowers have five petals, five sepals, five stamens, and a pistil.
4. **Tubers**: The most economically important part of the potato plant is the tuber. Tubers are swollen underground stems that store starch and serve as the primary means of vegetative propagation.

Floral Biology

Potato flowers are hermaphroditic, containing both male and female reproductive organs. The flowers are typically self-fertilizing, but cross-pollination can occur with the help of insects. The fruit of the potato plant is a small, green, berry-like structure that contains numerous seeds. However, commercial potato production primarily relies on vegetative propagation through tubers rather than seeds.

1. **Pollination**: Potato flowers are usually pollinated by bees and other insects. Pollination is essential for seed production, although it is not commonly used for commercial propagation.
2. **Fertilization**: After pollination, fertilization occurs within the ovary of the flower, leading to the development of seeds. However, as mentioned, tuber propagation is preferred for maintaining the genetic consistency of cultivars.
3. **Seed Formation**: The seeds formed within the berries are true seeds, but they are seldom used for planting due to genetic variability. Instead, tubers from the previous season are planted to produce new plants.

Commercial Potato Species

The commercial potato generally belongs to two species: *Solanum andigenum* and *Solanum tuberosum*.

Solanum andigenum

Solanum andigenum is not very common. The plants have thin and long stems with small and narrow leaflets. They produce flowers more profusely and have long stolons with mostly colored deep-eyed tubers.

Solanum tuberosum

Solanum tuberosum is more common. These plants have shorter and thicker stems, with larger and wider leaflets.

Resistance varieties of Potato Species for Biotic and Abiotic Stresses

In addition to these two species, several other important species exhibit resistance to various biotic and abiotic stresses. These species are listed in Table .

Species	**Utility**
Solanum demissum Lind (2n = 72)	Resistance to late blight, virus A, virus Y
S. stoloniferum (2n = 48)	Resistance to virus X, frost, Colorado beetle
S. vernei Bitt et Wittm (2n = 24)	Resistant to two species of *Heterodera* nematode
S. multidissectum Waek (2n = 24)	Resistant to two species of *Heterodera* nematode
S. antipovczii Bulk	Resistance to late blight
S. curtilobum Jug. et Buk	Resistant to frost
S. phureja Jug. et Buk	Non-dormant type used in breeding
S. chacoense Bitter	Non-dormant type used in breeding, tolerance to high temperature
S. anomalocalyx, *S. jamessi* Torr, *S. saltense*	Resistance to early blight

Classification Based on Origin and Day Length Response

Potatoes can also be divided based on their origin and response to day length, as shown in Table.

Distant Characters	***S. tuberosum***	***S. andigenum***
Origin	Andean region	Chile region
Day length response	Long day plants	Short day plants
Polyploidy	Many types of polyploids	Tetraploids
Distinct varieties	'Up-to-date', 'Magnum'	All the desi varieties like 'Darjeeling', 'Red Round', 'Phulwa', and 'Gola'

Classification of Potato Based on Crop Duration

Potatoes can be classified into early and late varieties based on crop duration, as shown in Table.

Character	Early Potato	Late Potato
Stolon length	Short	Long
Stem	Hollow	Solid
Petioles	Broader	Narrow
Stolon number	Fewer	More
Tuber color	White	Colored
Photoperiodism	Long day for tuber development	Short day for tuber development
Varieties	'Satha', 'Gola', 'Great Scot', 'Kufri Chandramukhi'	'Phulwa', 'Kufri Kissan', 'Kufri Safed'

Area and Production

In India, potatoes occupy approximately 20.79% of the total vegetable cultivation area, which is around 533,5447 hectares. In 1995-96, the area under potato cultivation was 1,109,000 hectares, with production reaching 18,843,300 metric tons, representing 26.32% of the total vegetable production. The productivity of potatoes in India was 16.99 MT/ha, with West Bengal achieving the highest productivity at 24.4 MT/ha (NHB, 1998). Potatoes are grown in almost all Indian states, with major producing states including Uttar Pradesh, West Bengal, Punjab, Karnataka, Assam, and Madhya Pradesh.

Major potato-producing belts in India are:

- » Himachal Pradesh: Shimla, Lahul Spiti, Mandi
- » Punjab: Jalandhar, Hoshiarpur, Ludhiana, Patiala
- » Haryana: Ambala, Kurukshetra, Hisar, Karnal
- » Uttar Pradesh: Farrukhabad, Etawah, Mainpuri, Barabanki, Allahabad, Badaun, Moradabad, Agra, Aligarh, Mathura, Faizabad
- » Madhya Pradesh: Sidhi, Satna, Rewa, Sagar, Rajgarh, Tikamgarh
- » Chhattisgarh: Sarguja, Raipur
- » Rajasthan: Jaipur, Kota, Udaipur
- » Gujarat: Khera, Dissa, Baransh Katha, Jamnagar, Baroda, Mehsana
- » Odisha: Cuttack, Dhenkanal, Puri, Sambalpur
- » West Bengal: Hooghly, Nadia, Bardhaman, North 24 Parganas
- » Maharashtra: Pune, Satara, Kolhapur, Nasik

- Karnataka: Belgaum, Dharwad, Hassan, Kolar
- Uttarakhand: Almora, Pilibhit, Haldwani
- Andhra Pradesh: Medak, Chittoor
- Tamil Nadu: Dindigul, Nilgiris
- Assam: Nagaon, Kamrup, Sonitpur

Globally, major potato-producing countries include China, Russia, Poland, the USA, Ukraine, Germany, the Netherlands, and the UK (NHB, 1998).

Agronomic Practices

Soil Preparation and Planting

A well-prepared soil provides sufficient room for tuber development and helps retain moisture. Fields are plowed to a depth of 20-35 cm, followed by 4-5 plowings with country plows or disc harrows. Clods must be broken to make the field well-pulverized and leveled.

Planting Time

The planting and harvesting times vary based on the region and season.

Region	Season	Planting Time	Harvesting Time
North Western Plains	Early Autumn	Mid-Sept	Mid Nov-Dec
	Autumn	Mid Oct	Jan
	Spring	Feb-March	April
North Central Plains	Winter	Mid-Oct	Feb-March

Seed Rate and Spacing

- Seed Rate: 25-35 q/ha
- Tubers of 40-50 g with 40-50 mm diameter are planted at a spacing of 45-60 cm between rows and 20-25 cm within the rows. Large tubers are cut into pieces, each containing at least 1-2 eyes. Tuber cutting is not recommended for seed crop production due to the risk of transmitting viruses and bacteria.

Methods of Planting

The ridge and furrow method is popular, either manually or mechanically. Seed tubers should not contact fertilizers directly. Mechanical methods involve using tractor-drawn markers and fertilizer drills for furrowing, followed by planting with a planter-cum-ridger.

Manures and Fertilizers

Apply farmyard manure at 100 q/ha during field preparation. The general fertilizer recommendation is 120:80:60 kg N:P2O5 per hectare. Full doses of farmyard manure, phosphorus, and potassium, and half of the nitrogen are applied at planting. The remaining nitrogen is top-dressed during earthing up.

Intercultural and Weed Control

Mulching helps conserve soil moisture and reduce soil temperature. Weeds are controlled by cultural, chemical, or combined methods. Cultural methods include hoeing and weeding, followed by earthing up. Pre-emergence herbicides such as fluchloralin, alachlor, pendimethalin, and atrazine are effective. Post-emergence application of paraquat is also beneficial.

Irrigation

Pre-planting irrigation ensures uniform germination. Subsequent irrigations are given as required, with light and frequent applications preferred. Critical irrigation stages include stolon formation, tuber initiation, and tuber development. Irrigation is stopped about 10 days before harvesting to allow tuber skins to firm up.

Harvesting

The crop is harvested when fully matured, indicated by the yellowing of haulms and firm tuber skins. The field should be neither too wet nor too dry. Tractor-operated potato diggers are available for harvesting.

Yield

» Early varieties: 200 q/ha

» Late varieties: 300 q/ha

Physiological Disorders

1. **Hollow Heart:** Caused by rapid tuber growth, resulting in oversized tubers with cavities in the center. Maintain optimal soil moisture and avoid over-fertilization, especially nitrogen.
2. **Black Heart:** Results from sub-oxidation during storage due to poor aeration. Provide proper ventilation and avoid storing tubers in heaps.
3. **Greening:** Caused by exposure to sunlight, mechanical injury, or premature harvest. High glycoalkaloid content leads to solanin production. Proper earthing up and dark storage are preventive measures.
4. **Knobbiness:** Due to uneven growth conditions, particularly irregular watering. Ensure frequent and optimal irrigation.
5. **Cracking:** Caused by boron deficiency or uneven water supply. Apply Borax at 20 kg/ha and maintain regular irrigation.
6. **Sun Scalding:** Occurs in high temperatures and intense sunlight, leading to tip burn. Lower soil temperature by watering furrows.

7. **Black Spot:** Internal browning from mechanical injury, often linked to phenolic compounds. Provide proper storage and growing conditions.
8. **Freezing Injury:** Caused by exposure to freezing temperatures, resulting in tissue discoloration and necrosis. Avoid exposing tubers to freezing conditions.
9. **Sprouting:** A common storage issue, inhibited by spraying borax or iron sulfate before harvesting or using chemicals like Chloro IPC.

Table 1: Varieties of Potatoes

S. No	Variety Type	Varieties
1	Early Varieties	Kufri Ashoka, Kufri Chandermukhi, Kufri Jawahar, Kufri Lauvkar
2	Main Season Varieties	Kufri Jyoti, Kufri Sutlej, Kufri Pukhraj, Kufri Megha, Kufri Badshah, Kufri Anand, Kufri Bahar, Kufri Sadabahar, Kufri Deva, Kufri Sherpa, Kufri Swarna, Kufri Shailza, Kufri Surya, Kufri Himalini, Kufri Girdhari, Kufri Khyati
3	Late Varieties	Kufri Jeevan, Kufri Neelamani, Kufri Khasigaro, Kufri Naveen
4	Varieties for Processing	Kufri Chipsona 1, Kufri Chipsona 2, Kufri Chipsona 3, Kufri Himsona

Table 2: Agronomic Practices

S. No	Practice	Details
1	Soil Preparation	Plowing to 20-35 cm depth, breaking clods, leveling
2	Planting Time	Varies by region and season (refer to regional planting schedule)
3	Seed Rate and Spacing	25-35 q/ha, 40-50 g tubers at 45-60 cm row spacing, 20-25 cm intra-row spacing
4	Planting Methods	Ridge and furrow method, mechanical or manual
5	Manures and Fertilizers	100 q/ha farmyard manure, 120:80:60 kg N:P2O5 per ha
6	Interculture and Weed Control	Mulching, hoeing, weeding, pre and post-emergence herbicides
7	Irrigation	Light and frequent, critical stages: stolon formation, tuber initiation, tuber development
8	Harvesting	Fully matured crop, characterized by yellowing of haulms and firm tuber skins

Table 3: Physiological Disorders

S. No	Disorder	Cause and Management
1	Hollow Heart	Rapid growth, oversized tubers, optimal soil moisture, avoid over-fertilization
2	Black Heart	Poor aeration during storage, provide ventilation, avoid heaping
3	Greening	Sunlight exposure, proper earthing up, dark storage
4	Knobbiness	Uneven watering, frequent and optimal irrigation
5	Cracking	Boron deficiency, apply Borax, regular irrigation
6	Sun Scalding	High temperature and sunlight, water furrows to lower temperature
7	Black Spot	Mechanical injury, provide proper storage conditions
8	Freezing Injury	Freezing temperatures, avoid exposure during storage and harvest
9	Sprouting	Storage issue, borax or iron sulfate sprays, Chloro IPC

Importance and Uses

Potato is a staple food in many countries and has significant importance in feeding populations during emergencies due to its high yield and nutritional value. It is a vital source of starch and rich in carbohydrates, vitamins (B1, B2, B6, C), minerals (Ca, P, Fe), and proteins. Potatoes provide almost all dietary substances except fat, making them an essential part of the diet in many cultures.

Soil Requirements

Potatoes can grow in almost all soil types, but well-drained clay loam soil is considered ideal. On sandy loam soil, the crop can also be successfully grown with heavy manuring and proper irrigation. The optimal soil pH for potato cultivation is 6.0-6.5.

Climate Requirements

Potato is a cool-season crop that can tolerate moderate frost. The ideal soil temperature for germination is around 20°C. Young plant growth thrives at 24°C, but later growth is favored by a temperature of 18°C. Tuberization ceases when the night temperature exceeds 23°C, with the maximum tuberization occurring at 20°C. Tuber formation stops completely at temperatures around 29-30°C.

Varieties

Potato varieties are categorized into early, main season, and late varieties based on their maturity period:

Early Varieties

» Kufri Ashoka

- Kufri Chandermukhi
- Kufri Jawahar
- Kufri Lauvkar

Main Season Varieties

- Kufri Jyoti
- Kufri Sutlej
- Kufri Pukhraj
- Kufri Megha
- Kufri Badshah
- Kufri Anand
- Kufri Bahar
- Kufri Sadabahar
- Kufri Deva
- Kufri Sherpa
- Kufri Swarna
- Kufri Shailza
- Kufri Surya
- Kufri Himalini
- Kufri Girdhari
- Kufri Khyati

Late Varieties

- Kufri Jeevan
- Kufri Neelamani
- Kufri Khasigaro
- Kufri Naveen

Varieties for Processing

- Kufri Chipsona 1
- Kufri Chipsona 2
- Kufri Chipsona 3
- Kufri Himsona

Detailed Overview of Commercial Potato Varieties

KUFRI ALANKAR

Kufri Alankar is a potato variety resulting from the cross (Kennebee x O.N. 2090) x (Majestic x Ekishiraju). Released in 1968 by the Central Variety Release Committee, it is particularly suited for sandy soils in the plains of Punjab, Haryana, and Western Uttar Pradesh. The plants are tall, open, and erect with thick and strong stems that have a uniformly distributed color. The leaflets are rough, especially in the midrib of the top leaves, but not at the stem base. The tubers are oblong and taper towards the end, featuring dull white flesh and medium-deep eyes. This variety has a slow rate of degeneration and tends to develop cracks if allowed to mature fully. Kufri Alankar is known for its rapid bulking but has poor keeping quality. It matures in 75 days in the plains and 140 days in the hills, with a yield potential of 300 quintals per hectare (q/ha) in the plains and 100 q/ha in the hills. Kufri Alankar has moderate field resistance to late blight, immune to race 1 and '0', but it is susceptible to common scab.

KUFRI ANAND

Kufri Anand, derived from PJ376 x PH/F 1430, was released by the Central Potato Research Institute (CPRI), Shimla, in 1999. The plants are tall, erect, and vigorous with a few thick stems pigmented at the base. The foliage is grey-green, with intermediate leaves and a green rachis. The leaflets are ovate-lanceolate with a smooth, glossy surface and entire margins, occasionally displaying fused terminal leaflets. The flowers are light purple with moderate flowering. The tubers are white, large, oval-long, and flattened, featuring smooth skin and fleet eyes. The flesh is white, and the sprouts are red-purple. Kufri Anand matures in 100-110 days and yields between 350-400 q/ha. It is adaptable to the plains of Uttar Pradesh and neighboring states, resistant to late blight, and ideal for making French fries.

KUFRI ASHOKA

Kufri Ashoka is a versatile variety released from CPRI, Shimla, in 1996, resulting from the cross (EM/C-1020 x Allerfi'uii lleste Gelbe). The plants are medium tall, erect, medium compact, and vigorous, with few medium-thick stems lightly pigmented at the base. The foliage is green with intermediate leaves and a green rachis. The leaflets are ovate-lanceolate with a smooth, glossy surface and entire margins. The flowers are light red-purple with profuse flowering. The tubers are white, large, and oval-long, with smooth skin and fleet eyes. The flesh is white, and the sprouts are purple. Kufri Ashoka matures in 70-80 days, yielding 400 q/ha, and is suitable for cultivation in Bihar, Haryana, Punjab, Uttar Pradesh, and West Bengal. However, it is susceptible to late blight and not suitable for processing.

KUFRI BADSHAH

Kufri Badshah is a hybrid between Kufri Jyoti and Kufri Alankar, released in 1980 by the Central Subcommittee on Release of Varieties for the Indo-Gangetic plains of North India, including Punjab, Haryana, Uttar Pradesh, Bihar, West Bengal, Madhya Pradesh, and the plateau region. The plants are tall with erect haulms and have 4-5 stems per plant with a tendency for branching. The stems are semi-solid, flexible, slightly pubescent, red-purple with well-developed straight wings, and short internodes (about 7-8 cm). The leaves are green, medium-long, with prominent venation, and the rachis is pigmented at the base. The leaflets are broad, medium-sized, slightly hairy, with entire margins, ovate, and have prominent glassy veins. The tubers are mostly large to medium, oval, shining white, smooth, regular, with shallow eyes. The flesh is dull white, and the sprouts are stout, thick, and bulbous white with light red pigmentation at the base. The stolons are thick, short, and white. The flowers are scanty, white, medium-sized, with fully stretching wings, and pollen fertility is 50-60 percent. This variety has medium dormancy and average keeping quality. The tubers are tasty with a good flavor, containing 13.6 g/100 g starch and 18.6 percent dry matter content. It matures in 90-100 days in the plains and is tolerant to frost, resistant to late blight, early blight, and potato virus 'X', but susceptible to soft rot.

KUFRI BAHAR

Kufri Bahar is derived from the cross Kufri Red x Ginek, released in 1980 by the Central Subcommittee on Release of Varieties for the plains of Haryana, Punjab, and Western Uttar Pradesh. The plants are tall, erect, medium compact, and vigorous with thick stems (4-5 stems per plant) that are succulent, semi-solid, and have moderately developed wings. The leaves are open and straight, with 4-5 pairs of leaflets. The leaflets are grey-green, medium-sized, dull, with entire margins. The foliage is grey-green. The tubers are large, white, round to oval, with smooth skin and medium-deep eyes. The flesh is white, and the sprouts are bulbous and red. The stolons are medium-long. This variety has moderate flowering with white, medium-sized, sterile flowers. It has average keeping quality and medium dormancy. It matures in 90-100 days, with a yield of 250-300 q/ha. Kufri Bahar is resistant to late blight, early blight, and potato virus 'X', 'Y', and leaf roll, but it is susceptible to insect pests, drought, and frost.

KUFRI CHAMATKAR

Kufri Chamatkar is a derivative of the cross Ekishiraju x Phulwa, released in 1967 by the Central Variety Release Committee for the plains of Uttar Pradesh, Madhya Pradesh, Haryana, and Punjab, where one crop of long duration is grown. The plants are medium in size, and spreading, with short stolons, compact, dark green foliage, and prominent foliage. The internodes are very small, with more stems and small leaflets. The flowers are white. The tubers are round and white, with medium-deep eyes and yellow flesh. The sprouts are red. This variety shows slow degeneration and quick tuberization. It is a late-maturing variety, maturing in 110-120 days in the plains and 150 days in the hills, with a yield potential of 250 q/ha in the plains and 75 q/ha in the hills. Kufri Chamatkar is resistant to early blight but susceptible to viruses, late blight, brown rot, charcoal rot, wilts, and common scab.

KUFRI CHANDRAMUKHI

Kufri Chandramukhi is derived from the cross S. 4485 x Kufri Kuber and was released in 1967 by the Central Variety Release Committee. It is suitable for the plains of Punjab, Haryana, Uttar Pradesh, Madhya Pradesh, Rajasthan, Bihar, West Bengal, and Maharashtra, as well as the hills of Himachal Pradesh and the Jammu region. The plants are medium tall with open foliage, free from secondary growth. The foliage is grass green and glossy, with small leaflets that are not fully expanded, especially on the top, featuring a notched tip on one side. The tubers are oval and white with dull white flesh and fleet eyes. The sprouts are light red and pubescent. This variety has a slow rate of degeneration and early bulking, with good keeping quality. It matures in 80-90 days in the plains and 120 days in the hills, yielding an average of 200 q/ha in the plains and 75 q/ha in the hills. Kufri Chandramukhi is susceptible to common scab, late blight, brown rot, nematodes, charcoal rot, and wilts.

KUFRI CHIPSONA-1

Kufri Chipsona-1, a cross between MEX. 750826 and MS/78-79, was released from CPRI, Shimla, in 1998. The plants are medium tall, semi-erect, compact, and vigorous, with few, thick, slightly pigmented stems with poorly developed straight wings. The foliage is green, with open leaves and a green rachis. The leaflets are ovate, smooth, with a dull surface and entire margins. The flowers are white and profuse. The anthers are orange-yellow and well-developed, with a round pinhead stigma. The tubers are white, medium to large, oval, with smooth skin and fleet eyes, and dull white flesh with green sprouts. This variety matures in 90-110 days, yielding 400 q/ha. It is suitable for cultivation in Bihar and Uttar Pradesh and is resistant to late blight. Kufri Chipsona-1 is highly suitable for making chips and French fries.

KUFRI CHIPSONA-2

Kufri Chipsona-2 is derived from F-6 x QB/B-92-4 and was released from CPRI, Shimla, in 1998. The plants are medium tall, erect, open, and vigorous, with few, medium-thick, slightly pigmented stems with straight wings. The foliage is dark green, with open leaves and a green rachis. The leaflets are ovate-lanceolate, wrinkled, with a dull surface and slightly wavy margins. The flowers are white and moderate in flowering. The anthers are yellow and well-developed, with moderate pollen stability. The stigma is round and notched. The tubers are white, medium, round-oval, with smooth skin, fleet eyes, and yellow flesh with reddish-brown sprouts. It is a medium-maturing variety, yielding 350 q/ha. Kufri Chipsona-2 is resistant to late blight and tolerant to frost. Due to its high dry matter content, low reducing sugars, and low phenols, this variety is highly suitable for making chips and French fries.

KUFRI DEWA

Kufri Dewa is a derivative of the cross Craig's Defiance x Phulwa and was released by the Central Subcommittee on Release of Varieties in 1973 for the Tarai area of Uttaranchal and Shimla agro-climatic conditions. It is also suitable for Bihar and Orissa. The plants are very tall, open, and erect, with very small internodes. The leaves are semi-compact, dark green, and upright. The petioles are very long (3-6 cm towards the base) and devoid of leaves. The tubers are erratic round with purple splashes, deep and pigmented eyes, and light red sprouts with white flesh. The tubers develop brown spots after harvest. Kufri Dewa is highly susceptible to late blight but has good resistance to frost and drought. It matures in 120 days in the plains and 165 days in the hills, with an average yield of 250 q/ha.

KUFRI GIRIRAJ

Kufri Giriraj is suitable for both North and South India. It is a cross between SLB/1-132 x EX/A 680-16 and was released from CPRI, Shimla, in 1998. The plants are medium tall, semi-erect, medium compact, and vigorous, with many medium-thick stems colored at the base with moderately developed straight wings. The foliage is green, with open leaves and a green rachis. The leaflets are ovate, with a smooth, dull surface and entire margins. The flowers are light purple and shy flowering. The anthers are yellow, well-developed, with low pollen stability. The stigma is round. The tubers are white, medium to large, oval, with smooth skin, fleet eyes, and white flesh, with light purple sprouts. Kufri Giriraj has a yield potential of 250 q/ha. It is resistant to late blight (both foliage and tubers) but is not suitable for processing.

KUFRI HIMALINI

Kufri Himalini is a derivative of the cross SLB/H-140 x SLB/Z-389 (b), recommended by the 9th workshop of the All India Coordinated Potato Improvement Project for hilly regions, including the hills of Northern India and the Nilgiris in the South. The plants are tall, vigorous, erect, and semi-close, with a spreading tendency at maturity. The stems are 4-5, solid, slightly hairy, red-purple, spreading uniformly throughout the stem, with moderately developed straight wings. The leaves are semi-close, medium to long, with medium placement and a green rachis. The leaflets are grey-green, flat, with entire margins, and the petiole and midrib are green. The tubers are medium to large, oval, slightly flattened, with white skin, shallow to medium-deep eyes, dull white flesh, and a mealy texture. The sprouts are red-purple and medium thick. The flowers are profuse, persistent, white, with a light pinkish color on the tips of the corolla wings. The pollens are fertile, and natural berries are formed. The tubers possess medium dormancy (2-2.5 months) with good keeping quality, showing no shrinkage after five months of storage in the hills. Kufri Himalini is highly resistant to late blight but susceptible to early blight, phoma leaf spot, brown rot, common scab, and moderately susceptible to powdery scab and leaf blotch. Planting is done in March-April as a summer crop in Nilgiris in the northern hills, and August-September for the autumn crop in the southern hills. It matures 130-140 days after planting.

KUFRI JAWAHAR

Kufri Jawahar is derived from the cross Kufri Neelamani x Kufri Jyoti and was released from the Central Potato Research Institute, Shimla, in 1996. The plants are short, erect, compact, and vigorous, with few, thick stems lightly pigmented at the base and well-developed straight wings. The foliage is light green, with open leaves and a green rachis. The terminal leaflet is cordate, with a smooth, dull surface and entire margins. The flowers are white with moderate flowering. The anthers are orange-yellow, well-developed, with low pollen stability. The stigma is round and slightly notched. The tubers are creamy white, medium-sized, round-oval, with smooth skin, fleet eyes, and pale yellow flesh with purple sprouts. It is an early maturing variety, maturing in 80-90 days, with a yield potential of 400 q/ha. This variety is resistant to late blight and is not suitable for processing. Kufri Jawahar is widely adaptable in Haryana, Punjab, the plateau region of Gujarat, Karnataka, and Madhya Pradesh. It is also suitable for intensive cropping systems, offering flexibility for farmers in diverse climatic conditions.

KUFRI JEEVAN

Kufri Jeevan is a derivative of the cross M-I09-3 x D 698, adapted for the northwest hills of Himachal Pradesh and Uttar Pradesh. The tubers are medium-sized, with white skin, oval-shaped, fleet eyes, and faint red-pink, waxy, and pale yellow flesh. It matures in 150-160 days in the hills and 120 days in the plains. Kufri Jeevan is resistant to wart and highly resistant to late blight, with moderate resistance to early blight. This makes it a reliable choice for areas prone to these diseases, ensuring stable yields and reduced crop losses.

KUFRI JYOTI

Kufri Jyoti is a derivative of the cross 3069d(4) x 2814 Q(1), released in 1968 by the Central Variety Release Committee for Himachal Pradesh and Kumaon Hills of Uttaranchal, and also the plains where late blight is a limiting factor. The plants are tall, compact, and erect. The terminal leaflet is cottony or cordate, and the leaflets of the top 2-3 leaves are incurred at the base. The leaflets are broad and light green. The tubers are oval, white, with fleet eyes. The flesh is light white and waxy. The sprouts are blue-purple and medium thick. Kufri Jyoti has good keeping quality if given proper post-harvest treatment. It is field resistant to late blight with immunity to race. It is resistant to early blight and wart disease. It matures in 100 days in the plains and 120 days in the hills, with a yield potential of 200 q/ha in the plains and 150 q/ha in the hills. This variety is highly valued for its adaptability to various growing conditions and disease resistance.

KUFRI KHASHIGARO

Kufri Khashigaro is a popular variety in the hilly regions and is a derivative of the cross Taborky x SD 698 D. It is specifically adapted to the hilly regions of Assam. The tubers are medium, attractive white, round with deep eyes, and have waxy, pale yellow flesh with rapid bulking. This variety has a short dormancy period, allowing for quicker turnover between planting cycles. Kufri Khashigaro yields 160 q/ha in a 120-day crop duration. Its adaptability to hilly terrains makes it a reliable choice for farmers in these areas, offering consistent production and quality.

KUFRI LALIMA

Kufri Lalima is a fast-bulking variety and a derivative of the cross Kufri Red x CP 1362, released in 1982 by the Central Subcommittee on Release of Varieties for the plains of

Uttar Pradesh, Bihar, West Bengal, Orissa, and Karnataka. The plants are tall, vigorous, erect, compact, and dark green. The stems are 4-5, with red pigmentation prominent at the internodes. The leaves are open and wide, with red pigmentation running on the central side of the rachis. The leaflets are smooth, glossy, and ovate. The foliage is dark green. The tubers are uniformly red, medium to large, round, regular, smooth, with fleet to medium-deep eyes. The flesh is white with a mealy texture. The sprouts are medium thick and red. The flowers are numerous, persistent, and blue-purple. The pollens are fertile, with natural berry formation. Kufri Lalima is susceptible to late blight, wart, powdery scab, leaf blotch, phoma leaf spot, brown rot, charcoal rot, and common scab, and to virus 'Y'. It possesses a slow rate of degeneration and medium dormancy (about 2.5 months). This variety is favored for its attractive red tubers and adaptability to various agro-climatic conditions.

KUFRI LAUVKAR

Kufri Lauvkar is a derivative of the cross Serkoy x Adina, released in 1973 by the Central Subcommittee on Release of Varieties for the Deccan Peninsula (Maharashtra). The plants are tall, erect, with thick stems, large ovate-lanceolate leaflets, and dark grey-green foliage. The foliage is abundant, large, and mostly borne on the rachis. The tubers are round, white, with white flesh, and beetroot-red sprouts. The cooked flesh is white and good in taste. Kufri Lauvkar has a slow rate of degeneration. It is susceptible to late blight, leaf blotch, potato virus 'Y', and leaf roll but escapes attack by tuber moths. It is also susceptible to early blight. This early-maturing variety yields 250 q/ha in a 75-95 day crop duration. Its early maturity allows for flexibility in planting schedules, making it suitable for regions with short growing seasons.

KUFRI MUTHU

Kufri Muthu is a derivative of the cross 3046(1) x M-I09-C, released in 1971 by the Central Subcommittee on Release of Varieties for the Nilgiri Hills for summer and autumn seasons. The plants are tall and vigorous, with dark green leaves and short stolons. The tubers are large, white, roundish oval, with smooth skin, medium-deep eyes, and white flesh. The sprouts are blue-purple. Kufri Muthu has a yield potential of 135 q/ha in a 100-day crop duration. It is resistant to late blight but susceptible to early blight. This variety is well-suited for the specific climatic conditions of the Nilgiri Hills, providing reliable yields during the summer and autumn seasons.

KUFRI NAVEEN

Kufri Naveen is a derivative of the cross 0-692 x 3070d (4) and is adapted to the northeast hills of Assam and the high altitude of Himachal Pradesh. The tubers are medium, white, and oval with fleet eyes. The flesh is waxy and pale yellow. Kufri Naveen is susceptible to early blight but resistant to wart. It possesses a high degree of field resistance/immunity to races 0, 1, and 4 of late blight. This makes it a valuable variety for regions prone to late blight, ensuring stable and healthy crops.

KUFRI PUKHRAJ

Kufri Pukhraj is a widely adaptable variety and a cross of Craig's Defiance x JEX/B-687, released in 1998 from CPRI, Shimla. The plants are tall, semi-erect, medium compact, and vigorous. The stems are few, medium-thick, green, with moderately developed straight wings. The foliage is dark grey-green, with closed, large-sized leaves and a green rachis. The leaflets are ovate to lanceolate, with a smooth, glossy surface and entire margins. The flowers are white, with moderate flowering. The anthers are orange-yellow, well-developed, with high pollen stability and a round stigma. The tubers are white, large, oval, slightly tapered, with smooth skin, fleet eyes, and yellow flesh with blue-purple sprouts. Kufri Pukhraj is an early-maturing variety, maturing in 70-90 days, with a yield potential of 400 q/ha. It is resistant to early blight and moderately resistant to late blight. This variety is suitable for Bihar, Gujarat, Haryana, Himachal Pradesh, Karnataka, Madhya Pradesh, Maharashtra, Orissa, Punjab, Uttar Pradesh, and West Bengal. Although not suitable for processing, its early maturity and disease resistance make it a popular choice among farmers.

KUFRI SHEETMAN

Kufri Sheetman is a derivative of the cross Craig Defiance x Phulwa, released in 1968 by the Central Variety Release Committee for the plains, especially frost-affected areas of Punjab, Rajasthan, Haryana, and Western Uttar Pradesh. The plants are tall, open, erect, and vigorous, with thick, firm, and erect stems. The leaflets are rough, dull green, with waxy margins. The tubers are oval and white, with fleet eyes. Kufri Sheetman is resistant to frost and drought but susceptible to late blight and leaf roll. This variety is specifically bred to withstand frost, making it suitable for regions prone to early or late frosts, ensuring crop survival and yield.

KUFRI SHERPA

Kufri Sherpa is a derivative of the cross Ultimus x Adina and is recommended for cultivation in the hills of West Bengal by the 9th Workshop of the All India Coordinated Potato Improvement Project. The plants are tall, vigorous, erect, semi-compact, with spreading pasture under wet conditions. The stems are few, highly branched, semi-solid, and slightly hairy. The leaflets are rough, hairy, oval, with prominent venation, yellow-green, with short, light green petioles, and medium-sized foliage. The tubers are medium-sized, round-flattened, with smooth, medium-thick, white skin, medium-deep eyes, and pale yellow flesh. The sprouts are conical, blue-purple, and medium thick. The flowers are scanty, persistent, pink, regular in shape, with pollen fertility of 50-60 percent. The tuber dormancy is medium (2-2.5 months), with good keeping quality. Kufri Sherpa matures in 135 days in the hills and 100 days in the plains, with a yield of 225-250 q/ha. It is susceptible to frost, brown rot, and leaf roll, but resistant to late blight (both foliage and tubers) and phoma leaf spot, moderately resistant to early blight, and highly immune to wart. This variety is particularly suited for hilly regions with wet conditions, providing reliable yields and disease resistance.

KUFRI SINDHURI

Kufri Sindhuri is a derivative of the cross Kufri Kundan x Kufri Red, released by the Central Variety Release Committee in 1966 for the plains of Punjab, Jammu, Orissa, Bihar, Haryana, Uttar Pradesh, Madhya Pradesh, and West Bengal. The plants are tall, erect, with thick stems having pigment in most parts of the stem. The leaf rachis is pigmented, with open and stiff leaves. The leaflets are dull, with unequal basal lobes and slightly pigmented petioles, but not the midrib. The tubers are round, light red, with medium-deep eyes and dull white flesh. The tubers are moderate in number, with firm skin not prone to brushing, and larger tubers have a tendency to develop hollow hearts. Kufri Sindhuri has good keeping quality. It is a late variety, maturing in 120 days in the plains and 165 days in the hills, with a yield of 300 q/ha in the plains. It is slightly resistant to frost, but susceptible to late blight, common scab, charcoal rot, and wilts, moderately resistant to early blight, and has field resistance to viral diseases. This variety is favored for its attractive red tubers and adaptability to various agro-climatic conditions.

KUFRI SUTLEJ

Kufri Sutlej is a derivative of Kufri Bahar x Kufri Alankar, released in 1996 from CPRI, Shimla. The plants are medium compact and vigorous, with few, thick, lightly pigmented stems at the base and moderately developed wavy wings. The foliage is grey-green, with ovate-lanceolate leaves and slightly pigmented rachis. The leaflets are ovate, with a smooth, dull surface and entire margins. The flowers are white with moderate flowering. The anthers are orange-yellow, well-developed, with high pollen stability, and the stigma is round and slightly notched. The tubers are white, large, oval, with smooth skin, fleet eyes, and white flesh with light red sprouts. Kufri Sutlej is a medium-maturing variety, maturing in 90-100 days, yielding 400 q/ha. This variety is moderately resistant to late blight and recommended for cultivation in Bihar, Haryana, Madhya Pradesh, Punjab, and Uttar Pradesh. It has good consumer quality due to its easy-to-cook, waxy texture, mild flavor, and lack of discoloration after cooking. It possesses medium dry matter content and is not suitable for processing.

KUFRI SWARNA

Kufri Swarna is a cross of Kufri Jyoti x (VIn) 2 (62.33.3), released in 1985 from CPRI, Shimla. The plants are tall, erect, compact, and vigorous, with few thick stems and prominent wings. The foliage is dark green, with open leaves, green rachis, and ovate leaflets that are moderately pubescent with entire margins. The flowers are white with profuse flowering. The anthers are orange-yellow, well-developed, with low pollen stability, and the stigma is round. The tubers are white, medium, round-oval, with smooth skin, fleet eyes, and white flesh with blue-purple sprouts. Kufri Swarna is widely adaptable for the hills of South India. It matures in 130-135 days in summer and 100-110 days in autumn. It is highly resistant to both species of cyst (*Globodera rostochinensis* and *G. pallida*) and resistant to early and late blight. This variety is not suitable for processing. Its resistance to multiple pests and diseases makes it a reliable choice for farmers in hilly regions, ensuring stable yields and reduced crop losses.

Climatic Requirements

Potato is primarily a cool-season crop, thriving from sea level to snow lines as long as there is sufficient moisture and fertile soil. In India, potatoes are typically grown in winter in the plains and as a summer crop in the northern hills. Although potato is a long-day plant, it is cultivated as a short-day plant. It requires favorable environmental conditions such as low temperatures and short days during tuberization for rapid bulking. The ideal temperature for tuber formation is around 20°C, and tuber formation decreases as the temperature rises. Tuberization is adversely affected at temperatures around 30°C. Higher temperatures increase the respiration rate, causing the carbohydrates produced by photosynthesis to be consumed rather than stored in the tuber. High temperatures at any stage of growth can affect leaflet size and reduce tuber formation. Long-day conditions with ample sunshine and cooler nights are essential for reducing disease spread. Additionally, adequate moisture throughout the growing season is crucial for maintaining consistent growth and tuber development, as water stress can lead to reduced yields and poor tuber quality.

Soil Conditions

Potatoes can be grown in a wide range of soils, from sandy loam to clay soil. The soil should be friable, well-aerated, fairly deep, and rich in organic matter. Well-drained sandy loam and medium loam soils are most suitable for potato cultivation. Soil structure and texture significantly affect tuber quality. Light soils are preferred because they promote uniform soil temperatures and make harvesting easier. Alkaline or saline soils are unsuitable for potato cultivation, while acidic soils (pH 5.0 to 6.5) are favorable as they limit scab diseases. Good soil drainage is also critical to prevent waterlogging, which can cause tuber rot and other diseases. Incorporating organic matter such as compost or well-rotted manure can improve soil structure, fertility, and moisture-holding capacity, supporting healthy plant growth and high yields.

Planting Time

Plains:

» Early Crop: Third week of September to first week of October.

» Main Crop: First week of October to third week of October.

» Late Crop: Third week of October to first week of November.

Hills:

» Planting time ranges from the third week of February to the second week of April.

» In the southern hills near Ootacamund in Nilgiris, planting is done three times a year: in February, April, and September.

» In the plateau regions of Maharashtra, Bihar, and Madhya Pradesh, potatoes are raised during the rainy and winter seasons.

- In the Mysore plateau, the summer crop is planted in April-June and the winter crop in October-December.

These planting times are designed to ensure that potatoes grow during periods with favorable temperatures and daylight conditions, maximizing tuber formation and yield.

Seed Rate, Methods of Sowing, and Spacing

The seed requirements per hectare based on seed size are as follows:

- Large size: 25-30 q/ha
- Medium size: 15-20 q/ha
- Small size: 10-15 q/ha
- Out tubers: 8-12 q/ha

Potatoes are mainly planted using two methods:

1. **Ridge and Furrow Method:**
   - Ridges are prepared based on the slope of the plot.
   - Potato tubers are planted in the furrows, which should not be too long to ensure convenient irrigation.
2. **Flat Bed Method:**
   - The plot is divided into beds of convenient length and width.
   - Shallow furrows are opened, and potato tubers are planted at recommended distances, covered with the original soil from the furrows.
   - Earthing is done when plants reach 10-12 cm in height.

Suitable plant spacing for potato seed grades:

- Diameter of Tuber (cm) | Planting Distance (cm)
  - 2.5-3.5: 50 x 20 or 60 x 15
  - 3.5-5.0: 60 x 25
  - 5.0-6.0: 60 x 40

Proper spacing ensures that each plant has enough room to grow and access nutrients, light, and water, which is essential for healthy development and high yields.

Nutritional Requirements and Management

Soils low in organic matter should be supplied with 250-500 q/ha of farmyard manure or compost during land preparation, preferably a fortnight before planting. Potato plants are heavy feeders and require:

- Nitrogen: 100-150 kg/ha

- Phosphorus: 80-100 kg/ha
- Potassium: 80-100 kg/ha

Two-thirds to three-fourths of the nitrogen, along with the whole quantity of phosphorus and potassium, should be applied at planting. The remaining nitrogen is applied 30-35 days after planting during the first earthing up or when plants reach 25-30 cm in height, either as top dressing or foliar feeding. Spraying essential micronutrients like boron, zinc, copper, iron, manganese, and molybdenum is done when deficiency symptoms appear. Proper nutrient management ensures that potato plants receive all the essential elements they need for robust growth, high yields, and good tuber quality. Regular soil testing can help determine the specific nutrient needs and deficiencies, allowing for precise fertilizer application and efficient use of resources.

Intercultural Operations

Weed Management

Weeds in potato crops include both broad-leaved and narrow-leaved types. Generally, the use of herbicides is unnecessary because earthing up destroys most weeds. However, hand pulling may be required for weeds growing on ridges. Pre-emergence application of nitrofen (1.0 kg a.i./ha) or alachlor (2.0 kg a.i./ha) and post-emergence application of propanil (1.0 kg a.i./ha) can be effective. Care should be taken to avoid contact with potato plants when spraying post-emergence herbicides. Effective weed management ensures that potatoes receive optimal nutrients, water, and sunlight, preventing competition and promoting healthy growth.

Earthing Up

Proper earthing up is crucial for tuber development, ensuring adequate aeration, moisture availability, and proper soil temperature. Earthing should be done when plants are 15-22 cm in height, typically during top dressing of nitrogenous fertilizers. The ridges should be high enough to cover the tubers, with a second earthing two weeks later if necessary. A mould board plough or ridger can be used for earthing up in large areas. This practice protects the tubers from sunlight, preventing greening and solanine formation, and helps in maintaining a loose soil structure for better root expansion.

Use of Plant Growth Regulators

Soaking potato seed tubers in CCC (500 mg/l), sodium ascorbate (100 mg/l), cytozyme (5%), or foliar sprays with ethephon (400 mg/l), CCC (25 mg/l), or garlic acid (10-100 mg/l) can increase tuber yield. Foliar sprays with mixtalol (1 or 2 mg/l) also result in higher tuber yields. These growth regulators enhance various physiological processes, such as cell division and elongation, leading to increased tuber size and number. The use of growth regulators must be carefully managed to avoid any adverse effects on plant health and yield quality.

Water Management

Before planting, ensure sufficient soil moisture for satisfactory sprouting. Light pre-irrigation or immediate post-planting irrigation may be necessary if soil moisture is insufficient. The rate of water use is low until 30-35 days after planting, making the first irrigation essential within this period. Further irrigation is done as needed, with the furrow method being the common irrigation practice for potatoes. Consistent moisture levels are crucial for uniform tuber development and to avoid issues such as cracking or deformities. Efficient water management practices, such as drip irrigation, can also help in conserving water and ensuring its optimal use.

Harvesting, Yield, and Storage

After harvesting, potatoes should be heaped under shade for a few days to harden the skin and separate the soil. Good crop management can yield 350-450 quintals of marketable potatoes per hectare. Sorting is crucial to remove cut, bruised, or insect- and disease-damaged tubers. Healthy tubers are graded based on diameter. Oversized tubers are in demand for chips, while very small tubers are sold for vegetable use. Both oversized and undersized tubers are unsuitable for seed purposes. Potatoes can be stored in cold storage at 4-7°C with appropriate relative humidity. Proper storage conditions extend the shelf life of potatoes and maintain their quality, ensuring a steady supply throughout the year.

Diseases and Pests of Potato

Non-Viral Diseases

1. **Early Blight (*Alternaria solani*)**
   - **Symptoms:** Infection begins on lower leaves with necrotic spots having concentric rings. The fungus survives in the soil and diseased plant debris. Tomato is a collateral host. High moisture and low-temperature favor disease development.
   - **Control Measures:**
     - Practice crop rotation.
     - Collect and burn plant debris after harvesting.
     - Spray the crop with Dithane M-45 at 0.2%, starting 30 to 35 days after planting and repeating every 10 to 15 days.
     - Grow early blight-tolerant varieties such as Kufri Naveen, Kufri Sindhuri, and Kufri Jeevan.
2. **Late Blight (*Phytophthora infestans*)**
   - **Symptoms:** Infection appears at the tips or edges of lower leaves with circular or irregular water-soaked spots. White downy fungus growth appears on the underside of the leaves around the spots. Cloudy weather facilitates rapid disease spread. Severe infection can cause rotting of all above-ground parts and tubers.

» **Control Measures:**

- Use only healthy, disease-free certified seed tubers.
- Spray the crop thoroughly with Dithane M-45 (2.0 kg/ha), Dithane Z-78 (2.5 kg/ha), or Difolatan (2.5 kg/ha) well in advance of disease appearance. Continue spraying at 5 to 6-day intervals during cloudy weather.
- Harvest tubers when foliage is completely dry or cleaned.
- Grow late blight-resistant varieties like Kufri Navaharl.
- Avoid excessive nitrogen and irrigation.

3. Black Scurf (*Rhizoctonia solani*)

» **Symptoms:** Infected plants may die, and stem canker may form. Affected plants may form aerial tubers. Black sclerotial bodies form on tubers. The disease is soil- and tuber-borne.

» **Control Measures:**

- Use certified seeds.
- Treat seed tubers with organo-mercurial fungicides containing 6% mercury (e.g., Agallol, Aretan, Emisan) for 5 to 10 minutes before planting and before storage.
- Dip tubers in a 1.75% solution of sulfuric acid for 20 minutes.
- Apply Brassicol at 30 kg per hectare at sowing.
- Apply sawdust at 25 q/ha with recommended nitrogen doses at least 15 days before sowing.

4. Common Scab of Potato (*Streptomyces scabies*)

» **Symptoms:** The pathogen survives in soil and infected tubers in cold storage and spreads through manure. Affected tubers show superficial roughened corky tissue areas.

» **Control Measures:**

- Use healthy, disease-free seed tubers.
- Disinfect tubers by dipping in mercurial fungicide suspension (e.g., Emisan-6 or Agallol-6) at 0.25% concentration for 5 minutes.
- Plant tubers shallow.
- Keep soil moist during tuberization.
- Practice crop rotation with non-host crops like beets and carrots.
- Maintain soil pH 5.0 to 5.3.

5. **Black Leg or Soft Rot (*Erwinia spp.*)**

» **Symptoms:** Caused by various bacterial species, infected plants develop a blackened, shriveled cortex at the base of shoots. Leaflets become reddish at the tips, branches stiffen, and haulms turn jet black at the soil level. Tubers become watery and emit an offensive sulfurous odor upon rotting.

» **Control Measures:**

- Use healthy seed tubers.
- Avoid planting cut seed potatoes.
- Avoid planting too early or too late in the season.
- Collect and destroy affected plants.
- Avoid injuring tubers during harvesting, transit, and storage.
- Wash tubers with chlorinated water before storage.
- Keep stores well-ventilated, dry, and cool.

6. **Bacterial Brown Rot or Wilt Disease (*Pseudomonas solanacearum*)**

» **Symptoms:** Affected plants show dwarfing and bronze discoloration of leaflets. Browning of xylem vessels is visible in affected stems or tubers, and a whitish bacterial ooze may emerge when squeezed.

» **Control Measures:**

- Use healthy, disease-free seed tubers.
- Collect and destroy infected plants.
- Avoid runoff or irrigation water from diseased fields.
- Disinfect cutting knives with a suitable bactericide solution.
- Practice crop rotation with non-host crops like maize, soybean, and red top grass for at least three years.
- Reduce organic manure application and apply high doses of inorganic fertilizers.
- Treat seed tubers in a 0.02% streptocycline solution for 30 minutes after giving a five mm deep cut.

7. **Charcoal Rot (*Macrophomina phaseolina*)**

» **Symptoms:** Affected plant roots turn brown, and rotting starts on the stem. The skin of the stem becomes ash-colored, and dark lesions appear at the lenticels of the tubers and around the stem end. Charcoal rot is common during hot and dry months in late-season crops.

» **Control Measures:**

- Use disease-free seed tubers.
- Grow early maturing varieties such as Kufri Alankar or Kufri Chandramukhi.
- Irrigate crops at regular intervals.
- Treat seed tubers with Agallol-6 or Aretan at a 0.25% solution before storage. Do not consume treated tubers.

8. Wart Disease of Potato (Synchytrium endobioticum**)**

» **Symptoms:** Caused by a fungus, affected plants show warty outgrowths on stems, stolons, and tubers. The wart consists of distorted, proliferated-branched structures growing together into a mass of hyperparasitic tissue.

» **Control Measures:**

- Avoid growing potatoes in wart-affected soil.
- Use disease-free seed tubers.
- Soil treatment with 5% formalin is effective but costly.
- Grow wart-tolerant varieties like Kufri Sherpa, Kufri Jyoti, Kufri Jeevan, and Kufri Muthu.

Viral Diseases

Leaf Roll

» **Symptoms:** Leaf roll disease in potatoes is caused by the potato leaf roll virus (*Potato virus 1*), *Solanum virus 14*, or *Corium solani* Holmes. This disease spreads through infected tubers and by aphids, notably *Myzus persicae*. Affected plants become dwarf and upright, with leaves that roll inwards, becoming thick and leathery. These changes in leaf texture and appearance hinder the plant's ability to photosynthesize efficiently, leading to reduced vigor and lower yield. Additionally, rolled leaves can exhibit a yellowish tint, signaling a decline in plant health.

Mosaic Disease

» **Types and Causal Viruses:**

- **Mild Mosaic:** Caused by *Potato Latent virus*, *Potato Virus X*, *Potato Mottle Virus*, *Solanum Virus-1*. This form of the disease leads to mild chlorosis and mottling on the leaves without severe deformation. The virus can reduce the photosynthetic efficiency of the plants, leading to moderate yield losses.
- **Rugose Mosaic:** Caused by *Potato Virus X* and *Potato Virus Y*. Symptoms include wrinkling and roughness of leaves along with mottling. Plants may also exhibit reduced growth and lower tuber quality.

- **Cinkle of Potato:** Caused by *Potato Virus X* and *Potato Virus A*. This type causes leaf crinkling and curling, as well as chlorotic patches on the foliage, significantly impairing the plant's ability to produce healthy tubers.

» **Control Measures:**

- **Plant Only Certified Seed Tubers:** Using certified seed tubers ensures that the planting material is free from viruses, reducing the risk of disease introduction and spread.
- **Apply Systemic Insecticides:** Use insecticides like Thimet, Temik, or Furadan in the furrow at 10 kg per hectare at planting to control aphid vectors that transmit the virus.
- **Uproot and Destroy Affected Plants:** Remove and destroy infected plants to prevent the spread of the virus to healthy plants.
- **Spray with Metasystox:** Applying Metasystox at 0.1% helps control the aphid population, reducing the transmission of the virus.
- **Remove Haulms Early:** Cut and remove haulms in early January to prevent virus transmission through seed potatoes, especially in regions where the virus is prevalent.

Phytoplasma Diseases

Potato Witches' Broom

» **Symptoms:** Infected plants develop numerous axillary and basal shoots, giving them a 'witches' broom' appearance. These shoots are typically small and pigmented, and the tubers attached to the infected plants are also small and may sprout prematurely. The disease is transmitted by grafting and leafhoppers, particularly *Orosius albicinctus*. The pathogen can infect various plants, including tomato, brinjal, *Datura fastuosa*, *D. metel*, *Nicandra physaloides*, *Vinca rosea*, and *Calendula officinalis*. This widespread host range complicates management and control measures.

» **Control:** Hot water treatment of affected tubers at 50°C for 10 minutes can effectively eliminate the pathogen, preventing symptoms for up to 6-7 weeks. This treatment helps ensure that the tubers are pathogen-free before planting.

Purple Top of Potato

» **Symptoms:** Affected plants exhibit purple pigmentation and rolling of the basal parts of young top leaflets. The plants become stunted with many axillary shoots and may develop aerial tubers. The root system is poorly developed, which further hampers the plant's growth. This disease is transmitted by grafting and leafhoppers such as *Alebroides nigrocutellatus*, *Orosius albicinctus*, and *Seriana equata*. Transmission via tubers can vary depending on the age of the plant at infection.

» **Control:** Treatment with oxytetracycline, benlate, and cycloheximide can show 70-96% remission of symptoms. These treatments help manage the disease by reducing the pathogen load in the plants.

Marginal Flavescence of Potato

» **Symptoms:** This disease causes flavescence or chlorosis along the margins of the upper leaves, leading to the development of small, narrow leaflets and stunted plant growth. Affected plants produce clusters of small tubers with shorter dormancy and hairy sprouts. The disease is transmitted by grafting and hoppers such as *Seriana equata*. These symptoms can significantly affect the plant's photosynthetic capability and overall productivity.

» **Control:** Treatment with antiamoebin combined with benlate or benlate plus BP can cause up to 80% remission of symptoms. These treatments are effective in reducing the severity of the disease and improving plant health.

Potato Phyllody

» **Symptoms:** Infected plants display extreme hairiness and roughening of leaves and stems. The leaves become chlorotic, and leaflets are small with enlarged petioles. The androecia are naked, and axillary aerial tubers may develop. Phyllod flowers are enlarged with cupped clays and corolla. These symptoms can severely deform the plant, affecting its growth and tuber production.

» **Control:** Hot water treatment at 50°C for 10-15 minutes can eliminate the pathogen, preventing the appearance of symptoms and ensuring healthier plants.

Pests

Cut Worms (*Agrotis spp.*, *Euxoa spp.*)

» **Symptoms:** These pests cut sprouts at ground level and feed at night. They also attack tubers, creating holes that reduce their market value. The damage caused by cut worms can lead to significant crop losses if not managed properly.

» **Control Measures:**

- **Spray with Dursban 20 EC:** Use 2.5 ml per liter of water or drench plants where damage is noticed.
- **Apply Phorate 10 G Granules:** Apply at 10 kg per hectare around plants and rake the soil to incorporate the granules.
- **Use Well-Rotted Farmyard Manure:** Incorporate well-rotted manure to improve soil health and reduce pest populations.

Epilachna Beetle

- **Symptoms:** Both adults and larvae feed on and skeletonize leaves, causing significant defoliation and reducing the plant's ability to photosynthesize effectively.
- **Control Measures:**
  - **Hand-Pick Eggs, Larvae, and Adults:** Physically remove these stages of the beetle to reduce their population.
  - **Spray with Carbaryl:** Apply at 0.1% concentration to control the beetle population.
  - **Dust with Sevin Dust:** Apply at 30 kg per hectare for effective control.

Leaf Eating Caterpillars (*Spilosoma obliqua, Spodoptera exigua*)

- **Symptoms:** These caterpillars feed on potato leaves, causing significant defoliation and reducing the plant's photosynthetic capacity.
- **Control Measures:**
  - **Spray with Thiodon 35 EC or Endocel:** Apply at 0.15% concentration to manage caterpillar populations.
  - **Create Borders with Carbaryl Dust:** Dust the field borders to prevent caterpillar invasion and control existing populations.

Aphids (*Myzus persicae*)

- **Symptoms:** Aphids suck sap from leaves, causing plants to weaken, leaves to yellow, and curl downwards. Honeydew secretion leads to sooty mold and other fungal diseases, further affecting plant health.
- **Control Measures:**
  - **Spray with Rogor, Metasystox, Nuvacron, or Monocil:** Apply at 1 ml per liter of water, repeating every 10-12 days.
  - **Apply Thimet 10 G Granules:** Apply at 10 kg per hectare in furrows at planting to control aphid populations.
  - **Cut Haulms Early:** Remove haulms in early January to prevent virus transmission through seed potatoes.

Jassids

- **Symptoms:** These slender, wedge-shaped green insects suck sap from the underside of leaves, causing yellowing, curling, and later burning of leaves, known as 'hopper burn.' This damage can significantly reduce the plant's photosynthetic efficiency and overall health.

» **Control Measures:**

- **Spray with Metasystox:** Apply at 0.1% concentration to manage jassid populations.
- **Apply Thimet 10 G Granules:** Use at 10 kg per hectare in furrows at planting to prevent infestations.

Potato Tuber Moth (*Phthorimaea operculella*)

» **Symptoms:** This pest is a major issue for potatoes in storage and also attacks plants in the field. It bores and tunnels into potato tubers, causing significant damage that can reduce their market value.

» **Control Measures:**

- **Sow Healthy, Insect-Free Tubers:** Start with clean planting material to reduce the risk of infestation.
- **Use Well-Rotted Farmyard Manure:** Improve soil health and reduce pest populations.
- **Earth Up Carefully:** Ensure tubers are covered to prevent exposure to ovipositing female moths.
- **Spray with Carbaryl, Nuvacron, or Monocil:** Apply at 0.1% concentration to control moth populations.
- **Disinfect Storage Areas:** Use Malathion at 0.05% before storing potatoes to prevent infestations.
- **Sort and Remove Affected Tubers:** Remove damaged tubers before storage to minimize the risk of spread.

Golden Nematode (*Globodera rostochiensis, Heterodera rostochiensis*)

» **Symptoms:** Also known as potato root eelworm or potato cyst nematode, infected plants produce additional rootlets near the soil surface. Outer leaves become prematurely yellow and die, leading to reduced plant vigor and yield.

» **Control Measures:**

- **Quarantine Infested Areas:** Prevent the spread of nematodes by restricting movement from infested fields.
- **Avoid Growing Susceptible Crops:** Do not grow tomato or brinjal in infested soil to break the life cycle of the nematodes.
- **Follow Crop Rotation:** Exclude solanaceous crops and rotate with non-host crops to reduce nematode populations.
- **Treat Soil with DD:** Apply 90 gallons per hectare to manage nematode populations.

- **Avoid Soil or Seed Tubers from Infested Areas:** Use clean planting material and soil to prevent introduction of nematodes.

Root Knot Nematodes (*Meloidogyne incognita* and *M. javanica*)

» **Symptoms:** Affected plants are stunted and sickly, showing premature drying. Tiny galls form on tubers, reducing their market value. Nematodes survive as egg masses in diseased roots and tubers in soil, making management challenging.

» **Control Measures:** Implement similar practices as described for tomatoes, including crop rotation, soil treatment, and use of resistant varieties. Proper soil management and sanitation practices can help minimize nematode infestations and their impact on potato crops.

2. Tomato (*Solanum lycopersicum*)

Tomatoes (*Solanum lycopersicum*) possess a chromosome number of 2n=24. They originate from the Andean region of South America, particularly modern-day Peru. This region is considered the primary center of origin, where wild ancestors of the tomato were first domesticated by indigenous people. The secondary center of origin is Mexico, where the tomato was further cultivated and improved. The Aztecs were instrumental in the development of the tomato, utilizing it extensively in their cuisine. The Spanish Conquistadors introduced the tomato to Europe in the early 16th century, from where it spread globally.

Botanical Aspects

Tomato, scientifically known as *Solanum lycopersicum*, belongs to the Solanaceae family, which includes other economically important crops such as potatoes, eggplants, and peppers. It is a perennial plant in its native habitat but is typically grown as an annual in temperate regions. The plant is characterized by its pinnately lobed compound leaves and yellow flowers, which are typically self-pollinated. The fruit is a berry that comes in various shapes, sizes, and colors, ranging from small cherry tomatoes to large beefsteak varieties and from red to yellow, green, and even purple hues.

The tomato plant exhibits a sprawling growth habit, with stems that can be either determinate (bushy) or indeterminate (vining). The roots are fibrous and spread widely, aiding in nutrient and water absorption. Tomatoes are sensitive to frost and prefer warm weather, with optimal growth temperatures ranging from 70°F to 85°F (21°C to 29°C). They thrive in well-drained, fertile soils with a pH range of 6.0 to 6.8. The tomato plant's adaptability to various climates and soil types has contributed to its widespread cultivation.

Centers of Origin

The tomato has a rich history, originating in the Andean region of South America, particularly in modern-day Peru. This area is recognized as the primary center of origin for the tomato, where it was initially domesticated by indigenous people. The wild ancestors of the tomato were small, berry-like fruits that were likely consumed and selectively bred by early civilizations for desirable traits such as larger size and better taste. The secondary center of origin is Mexico, where the tomato was further cultivated and improved upon. The ancient Mexicans, including the Aztecs, played a significant role in the development of the tomato as we know it today. They utilized the fruit extensively in their cuisine, integrating it into various dishes. The Spanish Conquistadors are credited with introducing the tomato to Europe in the early 16th century, from where it eventually spread to other parts of the world.

Common Name: "Poor Man's Orange"

The tomato is often referred to as the "Poor Man's Orange" due to its affordability and high nutritional value. This nickname highlights its importance as a staple food for low-income populations, providing essential vitamins and minerals at a relatively low cost. Tomatoes are rich in vitamins A, C, and K, as well as potassium and folate. They also contain lycopene, a powerful antioxidant linked to various health benefits, including reduced risk of heart disease and certain cancers. The affordability and nutritional benefits of tomatoes make them a vital component of diets around the world, especially in regions where other nutrient-rich foods may be less accessible.

Area and Production

The cultivation of tomatoes has seen a significant expansion over the past four decades, reflecting its growing importance in agriculture and diet. In India, the area under tomato cultivation has surged from 36,000 hectares in 1960 to approximately 458,000 hectares today. Correspondingly, production has escalated to around 7.462 million tonnes, with a productivity rate of 16.29 tonnes per hectare. This remarkable growth can be attributed to advances in agricultural practices, improved varieties, and a greater awareness of the nutritional benefits of tomatoes. Leading tomato-producing states in India include Uttar Pradesh (UP), Karnataka, Maharashtra, Haryana, Punjab, and Bihar. These states have adopted intensive farming practices and have favorable climatic conditions that support tomato cultivation. The advancements in irrigation techniques, pest control, and fertilization have also played a crucial role in boosting tomato production.

Nutritive Value and Medicinal Use

Tomatoes are celebrated not only for their culinary versatility but also for their significant health benefits. They are low in calories, making them an excellent addition to a weight management diet. The high vitamin C content in tomatoes supports immune function and skin health. Vitamin A, present as beta-carotene, is crucial for vision and cellular communication. Potassium helps maintain normal blood pressure, and folate is essential for DNA synthesis and repair, making it particularly important during periods of rapid growth such as pregnancy. Lycopene, the compound responsible for the red color of tomatoes, is a potent antioxidant with several health benefits. Studies have shown that lycopene may help protect against certain types of cancer, particularly prostate cancer, and improve heart health by reducing levels of LDL cholesterol and inflammation.

Tomatoes are esteemed for their high nutritional value, occupying a prominent position among protective foods. They are rich in essential minerals, vitamins, and organic acids, which contribute to their health benefits. Key nutritional components of tomatoes include:

» **Minerals:**
- Calcium: 48 mg/100g
- Sodium: 12.9 mg/100g
- Copper: 0.19 mg/100g

» **Vitamins:**
- Vitamin A: 900 IU
- Vitamin C: 27 mg/100g
- Vitamin B complex (thiamine)

» **Organic Acids:**
- Citric acid
- Formic acid
- Acetic acid

The vibrant red color of tomatoes is due to the presence of lycopene, a powerful antioxidant, while the yellow color is attributed to carotenes. The unique flavor of tomatoes arises from compounds such as ethanol, acetaldehyde, and various volatile components. Tomatoes also contain tomatine, a steroidal glycoalkaloid found in different parts of the plant. Tomatine has been identified to have several health benefits, including antimicrobial properties and potential anti-cancer effects. Additionally, tomatoes are known for their culinary versatility and are used in various forms, including fresh, cooked, and processed products.

From a medicinal perspective, tomatoes are recognized for their appetite-stimulating properties. Tomato soup is often recommended as a remedy for preventing constipation due to its high fiber content and mild laxative effect.

Taxonomy

The botanical name of the tomato is *Solanum lycopersicum*, derived from the Greek words "lycos" (wolf) and "persicon" (peach). The genus *Lycopersicon* is divided into two subgenera:

1. **Eulycopersicon:** Includes red-fruited, self-compatible species.
   - *Lycopersicon esculentum* (now *Solanum lycopersicum*): Large-fruited varieties.
   - *Lycopersicon pimpinellifolium* (now *Solanum pimpinellifolium*): Small-fruited varieties.

2. **Eriolycopersicon:** Includes green-fruited, self-incompatible species.
    - *Lycopersicon hirsutum* (now *Solanum habrochaites*)
    - *Lycopersicon peruvianum* (now *Solanum peruvianum*)
    - Other wild species such as *Lycopersicon cheesmani* (now *Solanum cheesmani*)

Rick (1976) further categorized the genus *Lycopersicon* based on their ability to cross with the cultivated tomato:

» **Esculentum Complex (crossable with cultivated tomato):**
- *L. esculentum*, *L. pimpinellifolium*, *L. cheesemani*, *L. hirsutum*

» **Peruvianum Complex (not crossable with cultivated tomato):**
- *L. peruvianum*, *L. chinense*

Based on growth habit and fruit characteristics, five forms of *L. esculentum* are identified:

» **Communae:** Common tomato
» **Cerasiformae:** Small-fruited cherry tomato
» **Pyriformae:** Pear-shaped tomato
» **Grandifolium:** Potato-leaved tomato
» **Validum:** Erect and upright tomato

Botany

Tomato plants are true diploids with a chromosome number of 2n=24. They are typically annual plants with herbaceous, prostrate stems. The growth habit of tomatoes can be either determinate or indeterminate.

- **Determinate Growth:** The terminal bud ends in a floral bud, and further growth is arrested, resulting in a dwarf and bushy stature. This growth habit is often preferred for commercial production as it allows for a concentrated harvest period.
- **Indeterminate Growth:** The terminal bud remains a leafy bud, allowing continuous growth of terminal and lateral buds. This results in a more extended production of flowers and fruits along the main stem, often favored for home gardens and extended harvesting.

Tomato flowers are borne in racemose cymes, with flower clusters known as "trusses" positioned extra-axillary. The flowers are hermaphroditic, pendulous, pentamerous, and hypogenous. They have six stamens inserted on the throat of the corolla tube, with anthers converging around the style. Tomatoes are primarily self-pollinated due to the structure of their flowers, which include an introvert stigma, internal anther dehiscence, and synchronized stigma receptivity. Self-fertilization typically occurs as pollen is shed during the growth of the style through the anther cone. However, in warmer regions, some degree of cross-pollination may occur when the stigma protrudes beyond the anthers.

The sprawling growth habit of tomato plants, combined with their sensitivity to environmental conditions, makes them both a rewarding and challenging crop to cultivate. Proper management of temperature, soil conditions, and pollination can lead to high yields of high-quality fruits, making tomatoes a staple in gardens and farms worldwide.

Popular Varieties and Hybrids of Tomatoes

Tomatoes are widely grown in various regions, leading to the development of numerous varieties and hybrids. These are categorized based on their growth habits, with determinate and indeterminate types being the most common.

Determinate Varieties (Dwarf or Bush Type)

1. **Roma VF:** A popular determinate variety known for its pear-shaped fruits. It is well-suited for processing and cooking due to its meaty texture and rich flavor.
2. **Marglobe:** Another determinate variety with excellent disease resistance. It produces medium-sized, round fruits that are ideal for fresh consumption or canning.
3. **Celebrity:** Known for its high yields, disease resistance, and good flavor, Celebrity is a favorite among home gardeners and commercial growers alike.
4. **Rutgers:** A classic determinate variety with deep red fruits and rich, traditional tomato flavor. It is well-suited for slicing and canning.

Indeterminate Varieties (Tall Type)

1. **Better Boy:** An indeterminate hybrid known for its large, flavorful fruits. It is a popular choice for home gardeners due to its high yields and disease resistance.
2. **Brandywine:** A heirloom indeterminate variety prized for its exceptional flavor. It produces large, beefsteak-type fruits with a rich, sweet taste.
3. **Cherokee Purple:** Another heirloom variety with indeterminate growth. It produces large, dusky purple fruits with a complex, sweet flavor.
4. **Sun Gold:** An indeterminate cherry tomato variety with exceptionally sweet, golden-orange fruits. It is prized for its flavor and productivity.

S. No.	Variety	Special Features	Developing Institution	Recommended Areas
1	Arka Saurabh	Semi-determinate. Fruits firm, round, and nipple tipped.	IIHR, Bangalore	H.P., J&K, Hills of U.P., Punjab, Tarai region of U.P., Bihar, Jharkhand, Chhattisgarh, Odisha, A.P., Rajasthan, Gujarat, Haryana, Delhi, M.P., Maharashtra, Karnataka, Tamil Nadu and Kerala
2	Arka Vikas	Indeterminate, tolerant to moisture stress condition.	IIHR, Bangalore	H.P., J&K, Hills of U.P., Punjab, Tarai region of U.P., Bihar, Jharkhand, Chhattisgarh, Odisha, A.P., Rajasthan, Gujarat, Haryana, Delhi, M.P., Maharashtra, Karnataka, Tamil Nadu and Kerala
3	Arka Alok (BWR 5)	Bacterial wilt resistant, determinate. Fruits square and oblong.	IIHR, Bangalore	H.P., J&K, Hills of U.P., Punjab, Tarai region of U.P., Bihar, Jharkhand, Karnataka, Tamil Nadu and Kerala
4	Arka Vikas (Sel 22)	Determinate. Fruits medium large, oblate. Tolerant to mosaic.	IIHR, Bangalore	H.P., J&K, Hills of U.P., Punjab, Tarai region of U.P., Bihar, Jharkhand, Karnataka, Tamil Nadu and Kerala
5	Arka Saurabh (Sel.4)	Semi-determinate, resistant to fruit cracking, good transport quality. Fruits round firm with nipple tip.	IIHR, Bangalore	H.P., J&K, Hills of U.P., Punjab, Tarai region of U.P., Bihar, Jharkhand, Karnataka, Tamil Nadu and Kerala
6	Arka Ahuti (Sel.11)	Semi-determinate.	IIHR, Bangalore	H.P., J&K, Hills of U.P., Punjab, Tarai region of U.P., Bihar, Jharkhand, Karnataka, Tamil Nadu and Kerala

7	Arka Ashish	Tolerant to powdery mildew.	IIHR, Bangalore	H.P., J&K, Hills of U.P., Punjab, Tarai region of U.P., Bihar, Jharkhand, Karnataka, Tamil Nadu and Kerala
8	Arka Abha (BWR 1)	Resistant to bacterial wilt, semi-determinate.	IIHR, Bangalore	All over India
9	Arka Meghali	Suitable for rain fed cultivation.	IIHR, Bangalore	H.P., J&K, Hills of U.P., Punjab, Tarai region of U.P., Bihar, Jharkhand, Karnataka, Tamil Nadu and Kerala
10	SL 120	Resistant to nematode, determinate. Fruits flat round.	IARI, New Delhi	All over India
11	Pusa Early Dwarf	Determinate. Fruits flat oblate.	IARI, New Delhi	All over India
12	Pusa Uphar (DT 10)	Determinate. Fruits round with thick pericarp, suitable for processing.	IARI, New Delhi	Punjab, U.P., Bihar, Rajasthan, Gujarat, Haryana and Delhi
13	Sioux	Indeterminate. Fruits small and round.	IARI, New Delhi	All over India
14	Pusa Ruby	Indeterminate. Fruits flat round and suitable for processing.	IARI, New Delhi	All over India
15	Pusa Sadhabahar	Suitable for high and low temperature regimes.	IARI, New Delhi	NCR
16	Pusa Sheetal	Cold set variety (8°C or below). Fruits flat round.	IARI, New Delhi	Bihar, Punjab, Rajasthan, U.P., West Bengal, Delhi, Haryana, H.P. and J&K
17	Roma	Suitable for long distance transport, determinate growth.	IARI, New Delhi	All over India
18	Pusa Gaurav	Suitable for processing.	IARI, New Delhi	Punjab, U.P., Bihar, Rajasthan, Gujarat, Haryana and Delhi
19	Pusa Rohini	Longer shelf life. Fruits round.	IARI, New Delhi	NCR
20	H 24	Determinate, resistant to TLCV. Fruits flat round. Developed from cross between *L. esculentum* and *L. hirsutum L. glabratum*.	IIVR, Varanasi	J&K, H.P., Uttarakhand, Punjab, U.P., Bihar, Jharkhand, Chhattisgarh, Odisha, A.P., Karnataka, Tamil Nadu and Kerala

21	DVRT 2	Determinate. Developed from cross between *L. esculentum* and *L. hirsutum f. glabratum*. Fruits large and spherical.	IIVR, Varanasi	J&K, H.P., Uttarakhand, Punjab, U.P., Bihar, Jharkhand, Chhattisgarh, Odisha, A.P., Karnataka, Tamil Nadu and Kerala
22	Kashi Amrit (DVRT 1)	Developed from *L. esculentum* (*Sel.7*) and *L. hirsutum f. glabratum* through backcross pedigree selection. Fruits spherical with 108 g weight. Suitable for cultivation in TLCV infested period.	IIVR, Varanasi	U.P., Bihar and Jharkhand
23	La Bonita	Determinate. Fruits oblong with thick pericarp, suitable for long distance transport.	NBPGR, New Delhi	All over India
24	Mukthi (LE 79.5)	Bacterial wilt resistant, semi-determinate. Fruits round to flat round without green shoulder.	Kerala Agricultural University, Thrissur	All over India
25	Sakthi (LE 79)	Bacterial wilt resistant, semi-determinate. Fruits green shouldered and flat round.	Kerala Agricultural University, Thrissur	All over India
26	Anagha	Resistant to bacterial wilt and fruit cracking. Semi determinate, round medium sized fruits.	Kerala Agricultural University, Thrissur	All over India
27	Marutham (CO.3)	Developed by mutation breeding (EMS) from CO.1. Determinate. Fruits round and clustered.	Tamil Nadu Agricultural University, Coimbatore	All over India
28	CO.1	Semi determinate. Fruits round.	Tamil Nadu Agricultural University, Coimbatore	All over India
29	CO.2	Indeterminate. Fruits flat round and deep orange red.	Tamil Nadu Agricultural University, Coimbatore	All over India

30	PKM 1	Induced mutant suitable for long distance transport.	Tamil Nadu Agricultural University, Coimbatore	All over India
31	Paiyur 1	Suitable for rain fed cultivation. Less incidence of fruit borer. Fruits round, medium sized and slightly ribbing. Suitable for long distance transportation.	Tamil Nadu Agricultural University, Coimbatore	All over India
32	Pant T 3	Determinate. Fruits round, suitable for processing.	GBPUA&T, Pant Nagar	All over India
33	Pant Bahar	Resistant to Verticillium wilt and Fusarium wilt. Indeterminate. Fruits flat round and suitable for processing and storage.	GBPUA&T, Pant Nagar	All over India
34	Sel.32	Determinate. Fruits flat round and large.	Haryana Agricultural University, Hisar	All over India
35	HS 101	Determinate and dwarf. Fruits round.	Haryana Agricultural University, Hisar	All over India
36	Hisar Arun (Sel.7)	Determinate and dwarf. Fruits round.	Haryana Agricultural University, Hisar	All over India
37	HS 110	Potato leaved variety.	Haryana Agricultural University, Hisar	All over India
38	Hissar Lalima	Determinate. Fruits round and large.	Haryana Agricultural University, Hisar	All over India
39	Hisar Anmol	Resistant to leaf curl, determinate. Fruits flat round, medium size. Developed by crossing Hisar Arun and *L. hirsutum f. glabratum*.	Haryana Agricultural University, Hisar	All over India

40	Hisar Lalit	Resistant to root knot nematode. Determinate. Fruits medium large and round.	Haryana Agricultural University, Hisar	All over India
41	Punjab Chuhara	Determinate. Fruits pear-shaped and suitable for long distance transport, susceptible to TLCV.	Punjab Agricultural University, Ludhiana	Punjab, Haryana and Himachal Pradesh
42	Punjab Kesari	Determinate. Fruits pear-shaped and suitable for long distance transport, susceptible to TLCV.	Punjab Agricultural University, Ludhiana	Punjab, Haryana and Himachal Pradesh
43	Sel.1-6.1-4	Root knot nematode resistant, determinate, fruits pear-shaped.	Punjab Agricultural University, Ludhiana	All over India
44	Azad T2 (KS-2)	Determinate, early, fruits small round, moderately resistant to leaf curl virus.	CSAU&T, Kanpur	U.P., Bihar and Jharkhand
45	Uskal Kumar	Bacterial wilt resistant, fruits round and medium sized.	OUAT, Bhubaneswar	Odisha, Chhattisgarh, Andhra Pradesh, Telangana
46	Urkal Urbashi (BT12)	Bacterial wilt resistant, indeterminate, fruits pear-shaped and clustered.	OUAT, Bhubaneswar	Odisha, Chhattisgarh, Andhra Pradesh, Telangana
47	BT 18	Bacterial wilt resistant.	OUAT, Bhubaneswar	Odisha, Chhattisgarh, Andhra Pradesh, Telangana
48	BT 21	Bacterial wilt resistant.	OUAT, Bhubaneswar	Odisha, Chhattisgarh, Andhra Pradesh, Telangana
49	Narendra Tomato 1	Indeterminate, fruits large.	Narendra Dev Agricultural University, Faizabad	All over India
50	Narendra Tomato 2	Determinate, suitable for transport and processing. Resistant to nematode.	Narendra Dev Agricultural University, Faizabad	All over India

51	VC 48-1	Bacterial wilt resistant, determinate, fruits pear-shaped and clustered.	Assam Agricultural University, Jorhat	Assam, West Bengal, Odisha and North East States
52	Solan Gola	Round fruits.	YS Parmar University of Horticulture and Forestry, Solan	Himachal Pradesh, J&K, Uttarakhand
53	Swarna Naveen	Resistant to bacterial wilt. Fruits are medium sized (60 g), oblong, dark red with 5.0% TSS. Yield 60 t/ha in 120-140 days.	HARP, Ranchi	Jharkhand, Bihar, West Bengal, Odisha, Chhattisgarh, Uttar Pradesh
54	Pusa-120	Semi-determinate, spreading, late maturing, dark green foliage. Fruits flattish round, medium to large, uniform red, less acidic, less seeded, resistant to nematode, suitable for winter and summer seasons.	IARI, New Delhi	All over India
55	Pusa Ruby	Indeterminate, early. Fruits round, uniform ripening, good flavor.	IARI, New Delhi	All over India

Exploitation of Hybrid Vigour in Tomato Cultivation

Tomato (*Solanum lycopersicum*) serves as an exemplary case for the exploitation of hybrid vigor in vegetable cultivation. The increased consumer demand for tomatoes, coupled with efficient emasculation and pollination processes, higher seed yields per fruit, and the ability to combine numerous favorable genes in F1 hybrids, makes tomatoes ideal for heterosis breeding. Many F1 hybrids have been developed, exhibiting traits such as high yield, earliness, uniformity, superior quality, suitability for long-distance transport and storage, adaptability to adverse conditions, and resistance to biotic and abiotic stresses. Both public-sector research organizations and private seed companies have been instrumental in developing these hybrids. The success of hybrid tomatoes is largely due to their enhanced performance over traditional varieties, making them highly preferred by farmers worldwide.

Significant Tomato Hybrids

Following the introduction of the new seed policy by the Government of India in 1988, numerous private seed companies began engaging in vegetable research and F1 seed trade. As a result, F1 hybrids developed and marketed by private seed companies have become dominant in states like Maharashtra, Karnataka, Haryana, and Punjab. Below are some notable public-sector tomato hybrids:

Developing Institution	Variety	Special Features
IIHR, Bangalore	Arka Abhijit (BRH 1)	Determinate, resistant to bacterial wilt, 63.7 t/ha.
	Arka Vishal (FMH 1)	Indeterminate, fruit weight 140 g, round, green shouldered, resistant to cracking, good shelf life, 75 t/ha in 160 days.
	Arka Shresta (BRH 2)	Determinate, resistant to bacterial wilt, 68.5 t/ha.
	Arka Vardhan	Indeterminate, resistant to root knot nematode, oblate fruits, 75 t/ha in 160 days.
IARI, New Delhi	Pusa Hybrid 2	Semi-determinate, tolerant to nematode, 55 t/ha.
	Pusa Hybrid 1	Determinate, tolerant to high temperature (up to 23°C night temperature).
HARP, Ranchi	Swarna Baibhav	Fruits round, deep red, suitable for long-distance transportation, average yield 70-80 t/ha.
Tamil Nadu Agrl. University	COTH.1	Determinate. Fruits round to oblong, slightly acidic, 96 t/ha in 115 days.
	COTH 2	Semi-determinate, fruits in clusters of 4-5, high yield (90.2 t/ha), resistance to leaf curl virus.

Additional hybrids from private seed companies popular among farmers include Sun 496 (Sungrow Seeds, Delhi), HOE 303, Avinash 2 (Syngenta), BSS 20, Meenakshi, Tolstoi (Beejo Sheetal), NA 501, NA 601 (Nath Seeds), Sadabahar, Gulmohar, Sonali, MTH 6 (Mahyco, Jalna), ARTH 3, ARTH 4 (Ankur Seeds), Rashmi, Vaishali, Rupali, Naveen (IAHS, Bangalore), Swaraksha, Uttam (Namdhari Seeds, Bangalore), and Kt 4 (IARI, Katrain). These hybrids showcase the remarkable advantages of hybrid vigor in enhancing productivity and resilience in tomato cultivation.

Techniques for Hybrid Seed Production

Most commercial F1 hybrids are produced through hand emasculation and pollination, where female and male parent plants are planted in a 12:1 ratio. Flowers are typically emasculated in the afternoon and pollinated the next morning with freshly collected pollen. Pollen grains can also be stored for 2-3 days under normal conditions for pollination purposes. This method ensures precise pollen transfer from selected male parents to female flowers, maintaining the desired hybrid characteristics.

To reduce the cost of hybrid seed production, male sterile lines can be used. Male sterile plants in a population can be identified by their small-sized flowers, poorly developed anthers, lack of anthers, or anthers with sterile pollen grains. Once a male sterile plant is identified, it is stabilized for use as a female parent in commercial F1 seed production. Since male sterility is governed by recessive genes (msms), maintenance involves crossing with a heterozygous (Msms) fertile parent. The progeny segregates as male sterile (msms) and male fertile (Msms) plants in a 1:1 ratio. From this, male sterile plants are detected and used in the hybrid seed production program. This technique significantly enhances the efficiency and cost-effectiveness of hybrid seed production.

Optimal Climate for Tomato Cultivation

Tomato is a day-neutral, warm-season crop that cannot tolerate frost. Cool and dry weather is preferred, with optimal temperatures being 21-28°C during the day and 15-20°C at night. Night temperature is more critical than day temperature. High temperatures can result in exerted stigma, dryness of stigma, burning of anther tips, poor pollen dehiscence, low pollen viability, and slow pollen tube growth, leading to low pollination and fruit set. Viral disease incidence also increases at high temperatures. The optimal temperature for fruit color development is 21-24°C, and lycopene formation is hampered above 27°C. Seed and pollen germination are adversely affected below 10°C.

Based on night temperature requirements for fruit set, tomato varieties are classified into three categories:

» **Normal set varieties:** Set fruits at 15-20°C.

» **Hot set varieties:** Set fruits above 20°C (e.g., Philippine, Punjab Tropic, Pusa Hybrid 1).

» **Cold set varieties:** Set fruits below 15°C (e.g., Pusa Sheetal, Avalanche).

Tomatoes cannot withstand waterlogging, so well-drained, fertile soil rich in organic matter is preferred. The crop is moderately tolerant to acidic soil with a pH of 5.5, with the ideal pH being 6-7. The selection of appropriate varieties based on climatic conditions is crucial for maximizing yield and quality.

Sowing Time and Seed Rate

In regions with mild climates and no frost risk, three crops can be raised annually. In hilly areas, seeds are typically sown in March-April, while in plains, the crop is grown from June to November. In Kerala, seeds are sown in September and transplanted in October.

» Seed Rate:

- Open-pollinated variety: 400-500 g/ha
- Indeterminate F2 hybrid: 125-175 g/ha

- Seeds sown in an area of 200-240 m^2 (4-5 cents) are sufficient to plant one hectare. Four to five-week-old seedlings are used for transplanting. Seedling hardening, essential for their establishment in the main field, is achieved by withholding irrigation for a week before transplanting, adding NaCl (400 ppm) to irrigation water, or spraying cycocel (200 ppm) and zinc sulfate (0.25%) + 25 ppm proline at the time of transplanting. Proper seedling management ensures healthy and robust plants capable of high yields.

Field Preparation and Transplanting

Seedlings are transplanted on raised beds or ridges. The field is ploughed 4-5 times, and raised beds of 80-90 cm width or ridges and furrows are prepared. Spacing depends on the growth habit of the variety (determinate, indeterminate, or semi-determinate), with common spacings being 60 x 30-45 cm, 75 x 60 cm, and 75 x 75 cm. Closer spacing typically results in earlier and higher yields but may affect fruit size. Ensuring proper field preparation and transplanting techniques is vital for optimal plant growth and fruit production.

Fertilizer Application

Fertilizer recommendations for tomatoes vary by state and depend on the variety's growth habit and productivity. Typically, 15-20 tonnes of FYM, 100-125 kg N, 50-60 kg P_2O_5, and 50-60 kg K_2O are recommended per hectare. For F1 hybrids, a recommendation of 250:250:250 kg NPK/ha is common. FYM should be incorporated into the soil during the final ploughing. One-third of the N and the full doses of P and K are applied as a basal dose either just before transplanting or 5-10 days after transplanting. The remaining two-thirds of the N is applied at 20 and 45 days thereafter. In Tamil Nadu, an additional dose of 10 kg borax and 5 kg zinc sulfate as a basal dose is recommended to correct fruit cracking and increase yield and fruit quality. Proper fertilization ensures that plants receive the necessary nutrients for optimal growth and high yields.

Irrigation Practices

Furrow irrigation is the most common method for tomatoes, requiring adequate moisture throughout the growth period. Irrigation frequency depends on climatic and soil conditions, with summer crops needing irrigation every 3-4 days. Water stress during flowering can adversely affect fruiting and productivity. Long droughts followed by heavy irrigation can lead to fruit cracking, while a dry spell after regular irrigation can cause blossom end rot. Drip and sprinkler irrigation are becoming more common in areas with water shortages. Efficient water management practices are crucial to maintaining healthy plants and maximizing fruit production.

Weed Management

Frequent weeding, hoeing, and earthing up are essential to keep the field weed-free. Applying pendimethalin (1.0 kg a.i./ha) as a pre-emergence spray along with one hand weeding at

45 days after transplanting is ideal for the Pusa Ruby variety. Other effective herbicides include oxyfluorfen (0.25 kg a.i./ha), Goal (0.25 kg a.i./ha), and Basalin (1.0 kg a.i./ha) as pre-emergence applications. Post-emergence spray of Sencor (0.5 kg a.i./ha) is also effective. Mulching with straw or plastic can control weeds and regulate soil temperature. Effective weed management ensures that the plants do not compete for nutrients, water, and light, leading to healthier growth and higher yields.

Raising Tomato Seedlings

Nursery Bed Preparation and Seed Treatment

To cultivate tomato seedlings for one hectare, a net area of about 225 square meters is necessary. Nursery beds are generally 7.5 meters in length, 1 meter wide, and 10-15 centimeters high. It is essential to mix well-decomposed farmyard manure (FYM) at 3 kg per square meter into the topsoil. Additionally, incorporate a fertilizer mixture of 0.5 kg NPK (15:15:15) per bed into the soil at least 10 days before sowing seeds to enrich the soil with necessary nutrients.

Seed treatment is crucial for producing healthy seedlings. For one hectare, use 400-500 grams of seeds for open-pollinated varieties and 125-175 grams for hybrid varieties. Treat the seeds with fungicides such as Captan, Ceresan, or Thiram at a rate of 2 grams per kilogram of seed to protect against seed-borne diseases. The seed beds should also be treated with steam or 0.5 liters of 40% formalin per square meter of soil to sterilize the soil. After fumigation, cover the beds with polyethylene for 24 hours to enhance the efficacy. If fumigation is not feasible, employ solarization by covering the nursery bed with a transparent plastic sheet during the day for 10 days to reduce pest and disease attacks. In the absence of sterilization, drench the beds with 0.2% Brassicol or Captan solution to further protect the seedlings.

Sowing Methods and Irrigation

Seeds can be sown either by broadcasting or in rows, with rows spaced 7.5 centimeters apart. After sowing, cover the rows with a thin layer of compost and irrigate the beds using a rose can. Light watering should be done daily in the evening to maintain adequate moisture levels. Modern nurseries often employ low tunnels and low-cost polyhouses. Transparent plastic sheets used as low tunnels create an ideal microenvironment for seedling growth. Low-cost polyhouses enable winter planting for early spring transplanting, extending the growing season. Agronets can be used to protect seedlings from insect infestations and reduce vector-borne virus transmissions. Cover the beds with straw or polyethylene sheets until seeds germinate. Weekly fungicide sprays, such as Dithane M-45 or Difolatan at 0.25%, are recommended to minimize post-emergence damping-off.

Field Preparation and Transplanting

Preparing the Soil

For optimal growth, tomato fields should be well-prepared. Begin with an initial ploughing using a soil-turning plough, followed by 4-5 ploughings with a country plough. This helps to break down the soil and create a fine tilth. Level the field after ploughing to ensure uniformity. Transplant the seedlings onto raised beds that are 60-75 centimeters wide. Transplanting should be done in the late afternoon to minimize transplant shock, placing the seedlings on the sides of the beds to ensure they receive adequate moisture.

Spacing and Seed Rate

- **Open-Pollinated Varieties:** 400-500 grams of seed per hectare.
- **Hybrids:** 150-200 grams of seed per hectare.

Spacing for determinate varieties should be 60 centimeters between rows and 45 centimeters between plants. For indeterminate varieties, space 90 centimeters between rows and 30 centimeters within rows. Proper spacing ensures adequate air circulation and sunlight penetration, which are crucial for healthy plant growth.

Nutrient Management

Fertilizer Application

Tomatoes require a substantial amount of nutrients for optimal growth. During field preparation, incorporate well-rotted FYM at 200-250 quintals per hectare. Apply 75-100 kg of nitrogen (N), 50-75 kg of phosphorus (P_2O_5), and 50-60 kg of potassium (K_2O) per hectare. Apply one-third of the nitrogen, the full dose of phosphorus, and half of the potassium at planting. The remaining nitrogen should be applied one month after transplanting, and the rest of the potassium along with one-third of the nitrogen two months after transplanting.

Type	FYM (q/ha)	Nitrogen (kg/ha)	Phosphorus (kg/ha)	Potassium (kg/ha)
Open-pollinated	200-250	75-100	50-75	50-60
Hybrids	200-250	150-180	100-150	80-120

Proper nutrient management is vital to ensure that the plants receive all necessary nutrients for robust growth and high yields.

Weed Management and Intercultural Operations

Effective weed management is crucial in tomato fields, especially during the initial stages. The critical period for weed competition is 30-50 days after transplanting. Early weed control can be achieved using herbicides, followed by hand weeding in the later stages. Pre-emergence application of Alachlor (Lasso) at 2 kg active ingredient per hectare (4 liters per hectare in 750 liters of water) effectively controls annual and broad-leaved weeds. Pendimethalin (Stomp) at 1.2 kg active ingredient per hectare (4 liters per hectare) or Fluchloralin (Basalin) at 1.32 kg active ingredient per hectare (2.5 liters per hectare) can also be used for annual weeds. Consistent weeding and hoeing are necessary to keep the field clean and promote healthy plant growth.

Growth Hormones and Irrigation Practices

Hormone Application

- **GA_3 Sprays:** Apply GA_3 at 50 and 100 ppm on flower clusters and whole plants to promote fruit set and advance harvesting by one week. This hormone treatment encourages the development of more fruits and improves overall yield.
- **NAA Seedling Treatment:** Soak seedlings in NAA at 0.1 ppm for 24 hours in the dark to enhance fruit quality, set, earliness, and yield. This treatment improves the plants' resilience and productivity.

Irrigation Management

Careful irrigation is essential for tomato growth. Both over-watering and insufficient irrigation can be detrimental. Insufficient irrigation can arrest flower development, cause flower drop, and halt fruit growth. The most critical stages for irrigation are flowering and fruit development. Maintaining optimal soil moisture levels is crucial to prevent physiological disorders and ensure healthy plant growth.

Harvesting and Yield

Harvesting Stages

Tomato fruits are harvested at various maturity stages depending on their intended use and transportation distance. The recognized stages of maturity are:

1. **Immature Green Stage:** Green fruits of normal size with undeveloped seeds.
2. **Mature Green Stage:** Fully grown fruits with a brownish ring at the stem scar and a light green blossom end that changes to yellowish-green.
3. **Turning Stage (Breaker Stage):** 1/4th of the fruit, especially at the blossom end, shows a pink color.
4. **Pink Stage:** 3/4th of the surface shows a pink color.
5. **Hard Ripe Stage:** Nearly all red or pink with firm flesh.
6. **Over Ripe Stage:** Fully colored and soft, suitable for processing.

Yield

- **Open-Pollinated Varieties:** 250-300 quintals per hectare.
- **Hybrids:** 500-800 quintals per hectare.

Proper harvesting techniques and timing ensure that the fruits reach their market destination in the best possible condition, maximizing both yield and quality.

Managing Physiological Disorders in Tomato Plants

Physiological disorders in tomatoes can significantly impact crop yield and quality. These disorders result from abiotic stresses such as adverse weather conditions, nutritional deficiencies, or improper cultural practices. Effective management of these disorders is crucial for successful tomato cultivation.

1. Blotchy Ripening

Blotchy ripening is characterized by uneven ripening of fruits, with some portions showing color development while others remain greenish-yellow or whitish. This disorder is often caused by an imbalance of nitrogen and potassium in the soil, especially when potassium is deficient. To manage blotchy ripening, ensure balanced fertilization practices and proper irrigation to maintain optimal soil nutrient levels. Regular soil testing and appropriate fertilizer adjustments can help prevent this issue.

2. Blossom End Rot

Blossom end rot is a common disorder in tomatoes, characterized by the rotting of fruits at the blossom end. It is primarily caused by a deficiency of magnesium and calcium in the soil. To manage blossom end rot, it is recommended to spray calcium chloride at a concentration of 0.5% on the fruits during their development stage. This helps to correct the calcium deficiency and prevent the disorder. Additionally, maintaining consistent soil moisture levels can help mitigate the effects of this disorder.

3. Puffiness/Hollowness

Puffiness or hollowness in tomatoes refers to the development of light-weight fruits with hollow interiors. This disorder is caused by factors such as high or low temperatures, low soil temperature, and excessive soil moisture. To reduce puffiness, apply 4-CPA at a concentration of 20 mg/liter or CPPU at 20-25 mg/liter to the plants. These growth regulators help to improve fruit development and reduce the incidence of hollow fruits. Ensuring proper pollination and consistent environmental conditions can also help prevent this issue.

4. Sunscald

Sunscald occurs when tomatoes are exposed to excessive sunlight, particularly during extreme heat. This results in the development of white or grey-colored patches on the fruits. To protect tomatoes from sunscald, it is advisable to grow varieties with heavy foliage. The foliage provides shade to the fruits, reducing the risk of sun damage. Mulching and shading techniques can also be employed to protect the fruits from direct sunlight.

5. Cold Injury or Low-Temperature Injury

Tomatoes are sensitive to frost and can be severely damaged at near-freezing temperatures. Symptoms include withered vines and soft, water-soaked fruits. To protect tomatoes from cold injury, cover the plants with foliage or use frost covers to provide insulation against frost. Additionally, adjust planting schedules to avoid periods of frost. Growing cold-tolerant varieties can also help mitigate the risks associated with low temperatures.

6. Cracking of Fruits

Fruit cracking in tomatoes is a result of environmental factors such as boron deficiency and fluctuating moisture levels, particularly long dry spells followed by heavy watering. To manage this issue, apply 20-30 kg of borax per hectare to the soil to correct the boron deficiency. Additionally, ensure proper irrigation practices to maintain consistent soil moisture levels. Using mulch and proper irrigation scheduling can help prevent moisture fluctuations and reduce the incidence of fruit cracking.

3. Cole Crops

Cole crops, a group of vegetables that includes cauliflower, cabbage, broccoli, kohlrabi, kale, and Brussels sprouts, derive their name from the word "cole," an abbreviation of "caulis," meaning stem. These crops originated from a single wild ancestor, *Brassica oleracea* var. *oleracea* (sylvestris), commonly known as wild cabbage. Cole crops are widely cultivated during the winter season, with cauliflower and cabbage being the most significant. Broccoli is also gaining popularity due to its high medicinal value.

Origin and History

The cole crops have a rich history, tracing their origin to the Mediterranean region where *Brassica oleracea* var. *sylvestris* grew wild. Over time, various types of cole crops have been selectively bred from this ancestor, resulting in a wide range of vegetables that are now cultivated globally.

Nutritional Value and Uses

Cole crops are valued for their high nutritional content and health benefits. They are rich in vitamins, minerals, and antioxidants, which contribute to their medicinal properties. For instance, broccoli is renowned for its high levels of vitamin C, vitamin K, and folate, as well as its potential anticancer properties due to the presence of sulforaphane.

Area and Production

Cole crops are grown in various parts of the world, with significant production in regions with cool, moist climates. In India, these crops are primarily cultivated during the winter season. The table below provides an overview of the area, production, and productivity of cole crops in India:

Crop	Area (000' HA)	Production (000' MT)	Productivity (MT/HA)
Cauliflower	235.4	5824.1	24.7
Cabbage	239.9	5624.8	23.4
Broccoli	10.2	150.3	14.7

Popular Varieties and Hybrids

Cauliflower

- **Pusa Snowball K-1:** A popular variety known for its compact, snow-white curds. It is suitable for cultivation in cool climates.
- **Snowball 16:** This variety has a good market acceptance due to its attractive curd and high yield.

Cabbage

- **Golden Acre:** An early maturing variety with solid heads, suitable for fresh consumption.
- **Pusa Drum Head:** Known for its large, flat heads, this variety is excellent for sauerkraut and storage.

Broccoli

- **Pusa Broccoli KTS-1:** A high-yielding variety with large, green heads. It has good tolerance to pests and diseases.
- **Green Head:** Known for its tight, dome-shaped heads, this variety is popular in both fresh and processed markets.

Cultivation Practices for Cole Crops

Soil Preparation

Cole crops thrive in well-drained, fertile soils with a pH of 6.0-7.5. The soil should be rich in organic matter. Before planting, incorporate well-decomposed farmyard manure at the rate of 25-30 tons per hectare to improve soil fertility and structure.

Sowing and Transplanting

- **Cauliflower and Cabbage:** Seeds are sown in nursery beds and transplanted 4-6 weeks later. The spacing varies with the variety; for instance, early cabbage varieties are planted at 45 x 45 cm, while late varieties are spaced at 60 x 60 cm.
- **Broccoli:** Direct seeding or transplanting can be used. For transplanting, seedlings are ready in about 4-6 weeks.

Fertilization

Adequate fertilization is crucial for the growth and yield of cole crops. A balanced application of nitrogen (N), phosphorus (P), and potassium (K) is recommended. Typically, 100-120 kg/ha of N, 60-80 kg/ha of P, and 60-80 kg/ha of K are applied. Split applications of nitrogen, with half applied at planting and the rest in two split doses, ensure continuous nutrient availability.

Irrigation

Cole crops require regular irrigation to maintain soil moisture, especially during critical growth stages like head formation in cabbage and curd development in cauliflower. Drip irrigation is recommended for efficient water use and better crop growth.

Weed Control

Weed control is essential to reduce competition for nutrients, water, and light. Manual weeding, along with the application of pre-emergence herbicides like Pendimethalin, can effectively manage weeds.

Pest and Disease Management

Cole crops are susceptible to various pests and diseases. Integrated pest management (IPM) practices, including the use of resistant varieties, biological control agents, and judicious use of chemical pesticides, are recommended. Common pests include aphids, diamondback moths, and cabbage loopers, while diseases like black rot, downy mildew, and clubroot can affect crop health.

Harvesting and Post-Harvest Management

Harvesting

The harvesting time for cole crops depends on the crop and variety. For instance, cauliflower is harvested when the curd is fully developed but still compact, while cabbage is harvested when the heads are firm. Broccoli should be harvested before the flower buds open.

Post-Harvest Handling

Post-harvest handling includes proper sorting, grading, and packing to maintain quality and extend shelf life. Cole crops can be stored under refrigerated conditions to reduce spoilage and maintain freshness.

Future Prospects and Conclusion

The demand for cole crops is expected to rise due to their nutritional and medicinal benefits. Improved varieties and hybrids with better yield, disease resistance, and adaptability to different climatic conditions will play a crucial role in meeting this demand. Sustainable cultivation practices, including integrated pest management and efficient water use, will enhance productivity and profitability for farmers.

a. Cabbage (*Brassica oleracea* var. *capitata* Linn)

Family: Cruciferae

Chromosome Number: 2n = 2x = 18

Origin and History

Cabbage, known scientifically as *Brassica oleracea var. capitata*, has its roots in a wild, non-heading type called 'Cole wart' (*Brassica oleracea var. sylvestris*). The genus *Brassica* comprises approximately 100 species, most of which are native to the Mediterranean region, and considered the primary center of origin for cabbage. Historical evidence suggests that the real-headed cabbage evolved in Germany. Meanwhile, Savoy cabbage originated in Italy and spread to France and Germany during the sixteenth and seventeenth centuries. Over time, cabbage cultivation expanded globally, and it is now widely grown in various regions, including Caribbean countries, Indonesia, Malaysia, and Central, East, and West America. The adaptability of cabbage to different climates and its nutritional benefits have contributed to its widespread cultivation.

Nutritive Value and Uses

Cabbage is highly valued for its low-calorie content and rich nutritional profile. It contains only 27 calories per 100 grams, making it an excellent addition to weight management diets. Despite its low fat (0.1%) and carbohydrate (4.6%) content, cabbage is a good source of protein (1.3%), providing all essential amino acids, particularly sulfur-containing amino acids. The proteins in cabbage are comparable in biological value, digestibility, and net protein utilization (NPU) to those found in peas. Cabbage is rich in minerals, including calcium (39 mg), iron (0.8 mg), magnesium (10 mg), sodium (14.1 mg), potassium (114 mg), and phosphorus (44 mg) per 100 grams. It is also an excellent source of vitamins, including β-carotene (provitamin A), ascorbic acid (vitamin C), riboflavin (vitamin B2), niacin (vitamin B3), and thiamine (vitamin B1). The ascorbic acid content in cabbage ranges from 30-65 mg per 100 grams of fresh weight, although this can decrease significantly with cooking or shredding due to exposure to air.

Red cabbage, in particular, is notable for its high content of anthocyanin pigments, which include both mono- and diacylated cyanidin derivatives. These compounds contribute to the distinctive color and health benefits of red cabbage. Cabbage contains thermolabile sulfur compounds that produce sulfurous odors when cooked. Major volatile sulfur compounds, such as dimethyl disulfide and hydrogen sulfide, are derived from 1, S-methyl cysteine sulfoxide. Additionally, both white and red cabbage contain 2-propenyl isothiocyanate, and red cabbage contains 3-butenyl isothiocyanate, compounds known for their distinct flavors and potential health benefits.

Commonly used as a vegetable, cabbage is also popular in salads, often mixed with tomatoes, green chilies, and beetroot. The flavor of cabbage leaves is primarily due to the glycoside sinigrin. However, cabbage also contains goitrogens, which can cause enlargement of the thyroid gland if consumed in excessive amounts. Fortunately, most goitrogenic properties are lost during cooking. The thioglucosides in cabbage not only impart a desirable culinary flavor but also break down into products with anticancer properties. In Ayurvedic medicine, cabbage leaves are prescribed for various ailments, including cough, fever, skin diseases, peptic ulcers, urinary discharge, and hemorrhoids. Fresh cabbage juice has antibacterial properties and can inhibit the growth of various strains of lactic acid bacteria. Moreover, it is reported to have significant anticancer activity, further enhancing its medicinal value.

Historical Spread and Adaptation: Cabbage's journey from its wild ancestor to the cultivated varieties we know today highlights its adaptability and the role of human intervention in plant evolution. The spread of cabbage cultivation from the Mediterranean to Northern Europe and eventually to other parts of the world demonstrates its versatility and the ability to thrive in diverse climatic conditions.

Nutritional Comparison: When comparing the nutritional value of cabbage to other leafy vegetables, it stands out for its unique combination of low calories and high nutrient density. The presence of essential amino acids makes it a valuable protein source for vegetarians and vegans. Additionally, the rich mineral content supports various bodily functions, including bone health (calcium), oxygen transport (iron), and electrolyte balance (potassium).

Health Benefits Beyond Nutrition: Beyond its nutritional content, cabbage contains bioactive compounds that contribute to its health benefits. The presence of antioxidants such as anthocyanins in red cabbage helps combat oxidative stress, reducing the risk of chronic diseases. The sulfur compounds, while contributing to the distinct smell of cooked cabbage, also have detoxifying properties, aiding in liver function and overall detoxification processes in the body.

Culinary Versatility: Cabbage's versatility in the kitchen is another reason for its global popularity. It can be consumed raw in salads, fermented into sauerkraut or kimchi, cooked in soups and stews, or used as a wrap for various fillings. Each preparation method can influence the nutrient availability and health benefits, making it a versatile ingredient in diverse culinary traditions.

Medicinal Applications: The use of cabbage in traditional medicine underscores its therapeutic potential. Its application in treating gastrointestinal issues, such as peptic ulcers, highlights its soothing properties. The antibacterial activity of fresh cabbage juice offers a natural remedy for infections, while its anticancer properties provide a preventive approach to health management.

Sustainable Farming: Cabbage cultivation can also be part of sustainable farming practices. It can be grown in a variety of soil types and climates, making it accessible to farmers in different regions. Additionally, cabbage can be integrated into crop rotation systems to improve soil health and reduce pest and disease cycles, contributing to more sustainable agricultural practices.

Area and Production

India is the third-largest producer of cabbage globally, with major cultivation in states like Uttar Pradesh, Orissa, Bihar, Assam, West Bengal, Maharashtra, and Karnataka. The table below outlines the area, production, and productivity of cabbage in India from 1987-1988 to 1998-1999:

Year	Area (000' HA)	% of Total Area	Production (000' MT)	% of Total Production	Productivity (MT/HA)
1987-88	42.6	1.0	602.0	1.2	14.1
1991-92	177.3	3.2	2771.2	4.7	15.6
1992-93	188.9	3.7	3236.8	5.1	17.1
1993-94	231.7	4.8	3592.9	5.5	15.5
1994-95	216.8	4.3	3906.3	5.8	18.0
1995-96	218.4	4.1	3861.7	5.4	17.7
1996-97	210.2	3.8	3613.4	4.8	17.2
1997-98	228.2	4.1	5323.8	7.3	23.3
1998-99	239.9	4.1	5624.8	6.4	23.4

Productivity and Major Growing Regions of Cabbage in India

In the agricultural year 1998-99, India achieved a cabbage productivity of 23.4 tons per hectare, which was nearly on par with the global average of 24.1 tons per hectare. Tamil Nadu stood out with the highest productivity rates in the country. Key regions for cabbage cultivation in India include:

- **Uttarakhand**: This northern state, with areas like Almora, Chamoli, Uttarkashi, and Dehradun, benefits from its cool, temperate climate, making it ideal for cabbage cultivation. The favorable conditions here support the growth of high-quality cabbages, ensuring a steady supply to both local and national markets.
- **Uttar Pradesh Plains**: Known for their fertile soil, the plains of Uttar Pradesh provide an excellent environment for cabbage farming. The extensive agricultural practices in this region contribute significantly to the state's overall agricultural output, making it a major player in cabbage production.
- **Himachal Pradesh**: The hilly terrain and cooler climate of Himachal Pradesh are perfect for growing high-quality cabbages. The state's unique geographical conditions ensure that the cabbages produced here are of premium quality, catering to diverse market demands.
- **Haryana**: Regions like Hisar, Sonepat, Gurgaon, Ambala, Karnal, and Panipat are major contributors to Haryana's cabbage production. Advanced farming techniques

and efficient irrigation facilities in these areas support robust cabbage cultivation, enhancing both yield and quality.

» **Rajasthan**: Jaipur, Alwar, and Ganganagar are key areas where cabbage is grown. The state's efforts to improve horticultural practices and irrigation systems have significantly boosted cabbage production, making it an important crop in these regions.

» **Assam**: Known for its diverse agricultural practices, Assam includes cabbage as a significant crop due to its favorable climatic conditions and fertile soil. The state's emphasis on sustainable farming techniques further supports robust cabbage cultivation.

» **Bihar**: Important cabbage-growing districts in Bihar include Patna, Nalanda, West and East Champaran, Muzaffarpur, Vaishali, Bhagalpur, Darbhanga, and Madhubani. These regions rely heavily on cabbage farming, which plays an integral part in the local economy.

» **Gujarat**: Regions such as Mehsana, Kheda, Baroda, and Gandhinagar are known for their cabbage cultivation. Supported by the state's strong agricultural infrastructure, these areas contribute significantly to Gujarat's overall cabbage production.

» **Maharashtra**: Nasik and Pune are major hubs for cabbage production. The diverse climatic conditions in Maharashtra allow for extensive cabbage cultivation, ensuring a steady supply throughout the year.

» **Orissa**: Key areas like Baleshwar, Balangir, Cuttack, Dhankanan, Ganjam, Keonjhar, and Koraput are vital for cabbage farming. The state's efforts to enhance horticultural productivity have greatly supported the growth of cabbage in these regions.

» **Karnataka**: Belgaum, Hassan, Belvey, Mysore, and Dakshin Kannada are prominent regions for cabbage cultivation. Favorable weather and soil conditions in these areas contribute to high-quality cabbage production, meeting both local and regional demands.

» **Tamil Nadu**: The Nilgiris is a notable area for cabbage farming, benefiting from the cool climate and fertile soil. This region stands out as one of the top producers of cabbage in the country, known for its high yield and quality.

Globally, significant cabbage-producing countries include China, Russia, Japan, Korea, Poland, the USA, Indonesia, and Ukraine.

Classification of Cabbage

Bailey's Classification (1930) of Cultivated *Brassica oleracea*

Bailey classified *Brassica oleracea* into seven divisions based on their growth habits and morphological features:

A. Leaf Races (grown for their foliage):

1. **Branching shrubby kales** - *B. oleracea var. fruticosa*: These varieties are characterized by their branching growth habit and shrubby appearance. They are commonly grown for their nutritious leaves, which are used in various culinary dishes.
2. **Common biennial unbranching kales** - *B. oleracea var. acephala*: Known for their biennial growth cycle, these kales do not form heads and are cultivated for their broad, tender leaves. They are highly valued for their high nutritional content and versatility in cooking.
3. **Portuguese kales resembling cabbages** - *B. oleracea var. tronchuda*: These kales have a cabbage-like appearance and are valued for their large, fleshy leaves. They are used in traditional Portuguese dishes and are appreciated for their unique flavor and texture.

B. Head Races (grown for their compact heads or buds): 4. **Cabbage with terminal heads** - *B. oleracea var. capitata*: This category includes the traditional cabbage varieties that form tight, compact heads. These cabbages are widely used in salads, stir-fries, and soups due to their crisp texture and mild flavor. 5. **Brussels sprouts with small lateral heads** - *B. oleracea var. gemmifera*: Brussels sprouts are known for their small, edible buds that grow along the stem. Often roasted or steamed, they are popular for their unique flavor and high nutritional value.

C. Inflorescence Races (grown for their modified flower structures): 6. **Broccoli with partially modified flowers** - *B. oleracea var. italica*: Broccoli is cultivated for its thickened flower shoots, which are rich in vitamins and minerals. It is commonly used in a variety of dishes for its nutritional benefits and distinctive taste. 7. **Cauliflower with thickened and malformed flowers** - *B. oleracea var. botrytis*: Cauliflower is known for its dense, white flower heads. It is used in numerous culinary applications due to its mild flavor and versatility, making it a staple in many diets.

Cabbage can also be categorized based on various characteristics:

By Place of Origin:

- *B. oleracea var. capitata ssp. mediterranea* - Mediterranean cabbage: Originating from the Mediterranean region, these cabbages are adapted to warm climates and are often used in Mediterranean cuisine. They are known for their robust flavor and adaptability.
- *B. oleracea var. capitata ssp. orientalis* - Oriental cabbage: These varieties are popular in Asian cuisine and are known for their adaptability to diverse growing conditions. They are appreciated for their crisp texture and ability to enhance various dishes.
- *B. oleracea var. capitata ssp. europea* - European cabbage: Commonly found in Europe, these cabbages are well-suited to cooler climates. They are a staple in many European dishes, valued for their versatility and nutritional benefits.

By Botanical Color and Form of Heads:

» *B. oleracea var. capitata L.f. alba* - White cabbage: Known for its pale green or white heads, this type of cabbage is widely used in salads and coleslaw. It is appreciated for its crunchy texture and mild taste.

» *B. oleracea var. capitata L.f. rubra* - Red cabbage: Distinguished by its deep red or purple color, red cabbage is rich in antioxidants. It is often used in salads and pickling, adding a vibrant color and nutritional boost to dishes.

» *B. oleracea var. sabauda* - Savoy cabbage: Featuring crinkled, tender leaves, Savoy cabbage is prized for its mild flavor and versatility in cooking. It is commonly used in soups, stews, and as a wrap for various fillings.

By Head Shape (Choudhary, 1976):

» **Round head varieties**: Varieties like Golden Acre, Pride of India, Copenhagen Market, and Mammoth Rock Red are known for their perfectly round heads. These cabbages are ideal for various culinary uses, from salads to cooked dishes.

» **Flat head varieties**: Pusa Drum Head is a prominent example of flat head cabbage, known for its large, broad heads that are easy to slice and prepare. It is favored for its high yield and ease of use in the kitchen.

» **Conical head varieties**: Jersey Wakefield is a notable conical head variety, recognized for its pointed shape and early maturity. It is appreciated for its unique appearance and tender texture.

» **Savoy type varieties**: Chieftain is a popular Savoy type with crinkled leaves, offering a unique texture and flavor. It is used in various gourmet dishes for its visual appeal and delicate taste.

By Size and Shape of Head (Allen, 1914):

» **Wakefield and Winnings Kadt group**: This group includes varieties with small, conical heads that mature early, such as Jersey Wakefield and Charleston Wakefield. These cabbages are ideal for early harvests and fresh consumption.

» **Copenhagen Market group**: Known for their round heads and early maturity, varieties like Copenhagen Market, Golden Drumhead, Globe, and Bonanza are highly productive. They are favored for their uniformity and excellent flavor.

» **Flat Head or Drumhead group**: Featuring flattened heads, varieties like Pusa Drum Head, Early Round Dutch, All Head Early, Succession, and All Seasons are popular for their ease of preparation. These cabbages are versatile and used in a wide range of dishes.

» **Savoy group**: With wrinkled leaves and high quality, varieties like Drum Head

Savoy, Improved American Savoy, and Chieftain are grown for their unique texture and flavor. They are often used in gourmet cooking.

- **Danish Ball Head group**: These varieties, such as Danish Ball Head, Hollander, and Wisconsin Ball Head, are known for their thin leaves, compact heads, and excellent keeping quality. They are ideal for long-term storage.
- **Alpha group**: Characterized by very early, small, solid heads, varieties like Miniature Marrow are ideal for early harvests. They are appreciated for their quick growth and compact size.
- **Volga group**: Known for their thick, shiny blue leaves and loose bottom, varieties like Volga are distinctive and unique. They are favored for their attractive appearance and robust flavor.
- **Red Cabbage group**: Featuring red leaves similar to the Danish Ball Head, varieties like Red Rock, Red Dutch, and Large Red are valued for their color and nutritional benefits. They are often used in salads and pickling.

Popular Varieties and Hybrids

Determinate Type (Dwarf or Bush Type)

- **Golden Acre**: This early variety forms solid heads within 60-65 days, with each head weighing between 1-1.5 kg. It is popular for its quick maturity and compact size, making it ideal for fresh consumption and salads. The variety is also known for its resistance to splitting and excellent uniformity.
- **Pusa Drum Head**: Developed by the Indian Agricultural Research Institute (IARI), this variety produces flat, solid heads weighing 2-4 kg. It is resistant to black leg disease and is favored for its robustness and high yield. The heads are dense and heavy, making them suitable for storage and transportation.
- **Kranti**: Ready for harvest in about 60 days, Kranti produces compact heads weighing 0.8-1.2 kg. It is highly regarded for its early maturity and consistent performance, making it a reliable choice for farmers. The heads are firm and have a good shelf life, enhancing their marketability.

Indeterminate Type (Tall Type)

- **Jersey Wakefield**: Known for its early maturity, this variety forms pointed heads within 55-60 days, each weighing 1-1.5 kg. Its unique head shape and early yield make it a favorite among growers and consumers. The tender leaves and mild flavor make it suitable for a variety of culinary uses.
- **Pride of India**: Similar to Golden Acre but with larger heads weighing 1.5-2 kg, this variety is appreciated for its high yield and adaptability. The heads are dense

and have a long shelf life, making them ideal for both local and distant markets.

» **Late Drumhead**: This variety takes 105-110 days to mature, producing large, flat heads with a pale green color. It is ideal for late-season harvests and is known for its substantial size. The heads are robust and have excellent storage qualities, making them suitable for long-term use.

Nutritional Management

Cabbage requires specific manure and fertilizers depending on soil fertility. Generally, 200-250 q/ha of FYM is applied, and mixed thoroughly before transplanting. The quantity of nitrogen (N), phosphorus (P), and potassium (K) varies based on location. Apply half of the nitrogen and all phosphorus and potassium at transplanting, with the remaining nitrogen applied 30-45 days after.

State	N (kg/ha)	P (kg/ha)	K (kg/ha)
Madhya Pradesh	180	50	50
Uttar Pradesh	180	60	60
Bihar	150	-	-

Intercultural Operations

Regular intercultural operations, including manual weeding and hoeing, are necessary. Two to three manual weeding sessions are required. Herbicides like Trifluralin and Fluchloralin are effective for weed control.

Growth Hormones

Certain plant growth regulators can enhance growth and yield:

PGR	Concentration (mg/l)	Application Method	Attributes Affected
NAA	0.1	Seed treatment	Improves head size
IBA	0.4	Foliar spray	Increases yield
NAA + Chelated Zn	100 + 0.2 (percent)	Foliar spray (45 days after transplanting)	Increases yield

Water Management

Cabbage requires consistent soil moisture to achieve optimal growth and yield. Irrigate the plants immediately after transplanting to help them establish and reduce transplant shock. Subsequent irrigation should be done at intervals of 10-15 days, depending on the season and soil conditions. In hot and dry weather, more frequent watering might be necessary, while cooler and more humid conditions might extend the interval between irrigations. It's important to monitor the soil moisture levels regularly to avoid both drought stress and waterlogging. Water management is crucial during the heading stage; excessive irrigation at this point can lead to head splitting, a condition where the heads crack due to rapid water uptake. Mulching can also be employed to retain soil moisture, reduce temperature fluctuations, and suppress weed growth, contributing to better water management. Proper water management ensures that the plants remain healthy and produce high-quality heads, which is critical for achieving good market prices and reducing post-harvest losses.

Harvesting, Yield, and Storage

Harvesting cabbage at the right time is crucial to ensure maximum yield and quality. The timing of the harvest depends on the cultivar:

- **Early cultivars**: These take about 60-80 days from transplanting to reach marketable size. They are typically smaller but can be harvested sooner, providing early-season produce and a quicker return on investment for farmers.
- **Medium cultivars**: These take around 80-100 days to mature. They offer a balance between early and late varieties in terms of size and harvesting time, making them suitable for staggered planting and extended harvest periods.
- **Late cultivars**: These take approximately 100-130 days to mature. They are generally larger and are harvested later in the season, providing a continuous supply of cabbage and allowing farmers to meet market demands over a longer period.

Yields can vary based on the type of cultivar and growing conditions:

- **Early varieties**: Yields range from 300 to 400 quintals per hectare, which can be lucrative if marketed early when prices are higher.
- **Medium and late varieties**: Yields range from 400 to 600 quintals per hectare, with the potential for higher profits due to larger heads and extended market presence.

For storage, cabbage can be kept at a temperature of 0°C with a relative humidity of 90-95%, which can preserve the heads for 2 to 8 months. Proper storage conditions are essential to maintain the quality and extend the shelf life of the harvested cabbage, ensuring that farmers can supply the market consistently and reduce losses due to spoilage. Pre-cooling the cabbage before storage can also help in extending its shelf life by slowing down the respiration rate and maintaining freshness.

Physiological Disorders

Managing physiological disorders in cabbage is essential to ensure high-quality produce and reduce crop losses. Here are some common disorders and their management:

1. **Blossom End Rot**

- **Cause**: This disorder is primarily caused by a deficiency of magnesium and calcium in the soil, which disrupts the cell wall structure and leads to the decay of the fruit's blossom end.

» **Management:** To manage blossom end rot, spray calcium chloride at a concentration of 0.5% during the fruit development stage. Regularly monitoring and maintaining soil pH and nutrient levels can also help prevent this issue. Ensuring even moisture levels through consistent irrigation practices is crucial in minimizing the occurrence of this disorder.

2. **Cracking of Fruits**

» **Cause:** Fruit cracking is often due to boron deficiency and fluctuating moisture levels in the soil, which cause rapid growth and pressure on the fruit's skin.

» **Management:** Apply 20-30 kg of borax per hectare to correct the boron deficiency. Maintain consistent soil moisture levels to prevent cracking caused by rapid changes in water availability. Implementing proper irrigation scheduling and using mulch to conserve soil moisture can further help mitigate this issue.

3. **Puffiness/Hollowness**

» **Cause:** This disorder can result from extreme temperature fluctuations (either high or low), low soil temperature, or high soil moisture levels, which interfere with normal fruit development.

» **Management:** Apply 4-CPA at a concentration of 20 mg/liter or CPPU at 20-25 mg/liter to help reduce the incidence of hollow fruits and improve fruit quality. Ensuring optimal growing conditions, such as maintaining appropriate temperatures and avoiding over-irrigation, is essential in managing this disorder.

4. **Sunscald**

» **Cause:** Excessive sunlight exposure, particularly during extreme heat, can cause sunscald, leading to the development of white or grey patches on the fruits.

» **Management:** Grow varieties with heavy foliage to provide natural shade for the fruits, reducing the risk of sunscald. Additionally, using shade cloths during peak sunlight hours can protect the plants from excessive heat. Implementing proper spacing between plants can also help create a microclimate that protects the fruits from direct sunlight.

5. **Cold Injury**

» **Cause:** Near-freezing temperatures can cause cold injury to cabbage plants, resulting in withered vines and soft, water-soaked fruits, which are susceptible to secondary infections.

» **Management:** Adjust planting schedules to avoid periods of frost. Cover the plants with foliage or protective materials to provide insulation against cold temperatures. Utilizing frost protection methods, such as row covers or hoop houses, can also help mitigate the risk of cold injury.

6. **Blotchy Ripening**

» **Cause**: Blotchy ripening is often due to an imbalance of nitrogen and potassium in the soil, along with alternating sun and cloudy conditions that affect the uniform development of fruit color.

» **Management**: Ensure balanced fertilization practices to maintain optimal nutrient levels in the soil. Proper irrigation techniques help to manage this disorder by providing consistent moisture. Regular monitoring and adjusting of fertilization based on soil tests can prevent nutrient imbalances that lead to blotchy ripening.

Seed Production and Certification

Climatic Requirements

Cabbage thrives best in cool, moist climates with moderate rainfall. These conditions help in the formation of firm and dense heads. Cabbage plants can withstand frost during the head stage, making them suitable for regions with cold winters. For seed production, hilly areas with cool temperatures are ideal, as they provide the necessary conditions for vernalization—a period of cold required to induce flowering. Vernalization helps ensure the development of robust flower stalks, which are crucial for high-quality seed production. Consistent and appropriate climatic conditions during the growing season enhance seed quality and yield, making the choice of location critical for successful seed production.

Land Preparation and Transplanting

Proper land preparation is essential for optimal cabbage growth and seed production. The land should be free from volunteer plants that might carry diseases or pests. Preparing the soil to a fine tilth ensures good seedling establishment and root penetration. Incorporating 500-600 quintals per hectare of well-decomposed farmyard manure (FYM) into the soil improves soil structure and fertility. Transplant seedlings that are 3-4 weeks old during the main season (August to October) for late varieties. This timing allows the plants to grow and develop in favorable weather conditions, promoting healthy head formation and subsequent seed development.

Fertilizer Application

A balanced fertilization regime is vital for healthy cabbage growth and seed production. Apply 100 kg/ha of nitrogen, 60 kg/ha of phosphorus, and 60 kg/ha of potash. These nutrients support various growth stages, from vegetative growth to head formation and flowering. Additional nitrogen may be required before flowering starts to support the development of flower stalks and ensure robust seed production. Regular soil testing can help tailor fertilizer applications to meet specific soil nutrient needs, optimizing plant health and productivity.

Spacing and Irrigation

Proper spacing and irrigation practices are crucial for healthy cabbage plants and high seed yield:

- **Late varieties**: Space plants 60 x 60 cm apart to allow sufficient room for growth and air circulation.
- **Medium varieties**: Space plants 60 x 45 cm apart, balancing between sufficient space and efficient land use.
- **Early varieties**: Space plants 45 x 45 cm apart to accommodate their quicker growth cycle and smaller size.

Regular irrigation is essential, particularly during dry periods. However, avoid heavy irrigation during the head formation stage to prevent splitting and other water-related issues. Drip irrigation or furrow irrigation methods can be employed to ensure consistent moisture levels while minimizing water wastage.

Handling Mature Heads

For seed production, different methods can be used to handle mature cabbage heads:

- **Stump Method**: Remove the head, leaving a short stump with the roots intact.
- **Stump with Central Core Intact**: Retain the central core to support seed stalk growth.
- **Head Intact Method**: Leave the entire head on the plant for seed development.

During the rouging process, remove any off-type or diseased plants to maintain seed purity and quality. Consistent monitoring and removal of undesirable plants ensure that the remaining plants produce high-quality seeds that meet certification standards.

Harvesting and Threshing

Harvest the cabbage seed pods when 60-70% of them turn yellowish-brown, indicating maturity. Allow the harvested plants to cure for 4-5 days to ensure complete drying of the pods. Thresh the dried pods to separate the seeds, then further dry the seeds to reduce moisture content before storage. Proper drying prevents seed spoilage and maintains viability. The average seed yield for cabbage is 500-600 kg per hectare, depending on growing conditions and management practices.

Seed Certification Standards

Adhering to seed certification standards ensures the production of high-quality seeds that are true to type and free from diseases:

- **Field Inspections**: A minimum of three field inspections are required to monitor the crop for purity and health.
- **Field Isolation**: Maintain a distance of 1600 meters from other varieties and 1000 meters from non-conforming fields to prevent cross-pollination and contamination.

» **Field Standards**: Allow only 0.10-0.20% off-types and 0.10-0.50% seed-borne diseases in the field.

» **Seed Standards**: Ensure that the harvested seeds are 98% pure, with a maximum of 2% inert matter. The seeds should have a germination rate of at least 65% and a moisture content of no more than 7%.

b. Cauliflower (*Brassica oleracea* var. *botrytis*)

Family: Brassicaceae

Introduction

Cauliflower, a prominent member of the Cruciferae family, is scientifically classified as *Brassica oleracea* var. *botrytis*. Renowned for its edible white curd, cauliflower is highly nutritious and versatile, featuring prominently in various culinary applications worldwide. It is a resilient crop that can be grown under diverse environmental conditions, making it a dietary staple in many regions. The curd, which is the primary edible part, consists of undeveloped flower buds that are rich in nutrients.

Origin and History

The term "cauliflower" originates from the Latin words 'caulis' (cabbage) and 'floris' (flower), reflecting its close relationship with cabbage. Cauliflower traces its origins back to wild cabbage (*Brassica oleracea* var. *sylvestris*), known as 'Cole warts'. Through a process of natural mutation and selective breeding, humans developed the cauliflower we recognize today. Significant contributions to the development of modern cauliflower varieties came from European cultivars, particularly those from England, Germany, and the Netherlands.

In 1822, Dr. Jemson introduced cauliflower to India during the East India Company era (Swarup and Chatterjee, 1972). Initially, "Cornish types" from England were brought to India, followed by temperate types from Germany and the Netherlands in the 18th century. Indian cauliflower varieties emerged through inter-crossing European and Cornish types, resulting in tropical varieties that are resistant to high temperatures and heavy rainfall. This adaptability has facilitated the widespread cultivation of cauliflower across various climatic zones, from 11°N to 60°N latitudes globally.

Nutritive Value and Uses

Cauliflower's edible portion, known as the curd, comprises a shoot system with short internodes, branches, and bracts. Approximately 45% of the vegetable is edible, offering high-quality proteins and stable vitamin C content even after cooking. Cauliflower is an excellent source of dietary fiber, vitamins, and minerals, including potassium, sodium, iron, phosphorus, calcium, and magnesium. It also contains vitamins A, C, thiamine, riboflavin, and niacin.

A unique feature of cauliflower is its sulfur-containing compounds, which are responsible for its distinct flavor and potential health benefits. These compounds include hydrogen sulfide, methanethiol, ethanethiol, propanethiol, and dimethyl sulfide, which contribute to the characteristic odor when cauliflower is cooked. These sulfur compounds have been linked to various health benefits, including anti-carcinogenic properties.

Culinarily, cauliflower is extremely versatile. The curd can be used in an array of dishes such as curries, soups, and pickles. In regions with high production, the curd is often cut into pieces, dried, and preserved for off-season use. Cauliflower seedlings are also consumed in salads and as greens. The vegetable is reputed for its therapeutic effects, including treatment for scurvy, blood purification, and as an antacid. Cauliflower extract has been noted to inhibit carcinogenesis initiation and promotion, making it a valuable component of a healthy diet.

Area and Production

India stands as the largest producer of cauliflower globally, with substantial areas dedicated to its cultivation across various states. As of 1998-99, cauliflower occupied nearly 4.4% of the total 5335447 hectares of land under vegetable cultivation. During this period, cauliflower production was 4680600 tonnes. The area, production, and productivity of cauliflower have shown significant growth from 1987-88 to 1998-99, reflecting advancements in agricultural practices and variety development in Table 1.

Table 1: Area, Production, and Productivity of Cauliflower in India

Year	Area (000' HA)	Per cent of total area	Production (000' MT)	Per cent of total production	Productivity (MT/HA)
1987-88	75.1	1.8	558.60	1.10	7.4
1991-92	202.8	3.6	2998.10	5.10	14.8
1992-93	230.7	4.6	3612.20	5.70	15.7
1993-94	188.5	3.9	2872.50	4.40	15.2
1994-95	216.1	4.3	3244.10	4.80	15.0
1995-96	220.0	4.1	3244.10	4.80	15.0
1996-97	233.9	4.1	2474.00	3.50	11.20
1997-98	248.2	4.4	4471.00	6.20	18.00
1998-99	155.4	4.4	4680.60	5.30	18.40

Major Cauliflower Growing Belts

India

Cauliflower is extensively cultivated across various regions in India, each offering unique climatic conditions that support its growth. The major cauliflower growing belts in different states are as follows:

» **Uttarakhand**: The regions of Nainital, Pithaurgarh, and Dehradun are prominent for cauliflower cultivation. The cool, temperate climate of these hilly areas is ideal for growing high-quality cauliflowers.

» **Uttar Pradesh Plains**: The vast plains of Uttar Pradesh provide fertile soil and favorable weather conditions for extensive cauliflower farming.

» **Himachal Pradesh**: Shimla, Kullu, and Sirmour districts are known for their cauliflower production. The cooler climate and well-drained soils in these regions contribute to the successful cultivation of cauliflower.

» **Haryana**: This state has suitable climatic conditions and soil types that support robust cauliflower farming, making it one of the key producers in the country.

» **Rajasthan**: Known for its diverse agricultural practices, Rajasthan also contributes significantly to cauliflower production, particularly in regions with suitable soil and climate.

» **Bihar**: Several districts including Santhal Pargana, Purina, Katihar, Sitamarhi, Samastipur, Saharsa, Dhanbad, Patna, Nalanda, West and East Champaran, Muzzafarpur, Vaishali, Bhagalpur, Darbangha, and Madhubani are major cauliflower growing areas. The fertile plains and ample water supply in Bihar favor cauliflower cultivation.

» **Gujarat**: Districts like Mehsana, Khera, Ahmedabad, and Baroda are known for cauliflower farming. The state's irrigation facilities and suitable climatic conditions enhance cauliflower production.

» **Maharashtra**: Regions such as Nasik, Nagpur, and Kolhapur are prominent for cauliflower cultivation. The diverse climate and well-established agricultural infrastructure support the growth of this crop.

» **Orissa**: With its favorable weather and soil conditions, Orissa is also a significant producer of cauliflower in India.

» **Karnataka**: Districts like Hassan, Bangalore Rural and Urban, Kolar, and Mysore are known for their cauliflower production. The state's varied climate and agricultural practices contribute to successful cauliflower farming.

Global Producers

Beyond India, several other countries are notable for their significant cauliflower production:

» **China**: As the largest producer of cauliflower in the world, China has extensive cultivation areas and advanced agricultural practices that contribute to high yields.

» **France, Italy, and Spain**: These European countries have favorable climates and advanced farming techniques, making them major players in the global cauliflower market.

» **United Kingdom and Germany**: Both countries have a long history of vegetable cultivation, including cauliflower, with well-developed agricultural sectors.

» **United States**: The diverse climatic regions in the U.S. support extensive cauliflower farming, particularly in states like California.

» **Poland**: Known for its fertile soil and favorable growing conditions, Poland is a significant producer of cauliflower in Europe.

» **Pakistan**: With its suitable climate and growing demand, Pakistan is emerging as a notable producer of cauliflower in South Asia.

The widespread cultivation of cauliflower across these regions highlights its global importance as a nutritious and versatile vegetable. Each area's unique growing conditions and agricultural practices contribute to the diverse availability of cauliflower in markets worldwide.

Maturity Group	Nursery Sowing	Transplanting Time	Optimal Temperature Range for Curding	Varieties
Early I (A)	Mid-May to July beginning	Mid-July	20-25°C	Early Kunwari, Pant Gobhi-3, Pusa Meghna, Pusa Kartik Sankar
Early I (B)	May end to Mid-June	Mid-July	20-25°C	Pusa Katki, Pusa Deepali, Pant Gobhi-2
Mid Early (II)	July end to September beginning	Mid-July	16-20°C	Improved Japanese, Pusa Hybrid-2, Pusa Sharad, Pant Gobhi-4
Mid Late (III)	August end to September end	Mid-July	12-16°C	Pusa Synthetic, Pusa Subhra, Palam Uphar, Pant Subhra, Pusa HimJyoti, Pb Giant 35, Pusa Paushja, Pusa Shukti
Late (IV) Snowball	September end to mid-October	October end to mid-November	10-16°C	Snowball 16, Pusa Snowball-I, Pusa Snowball K-1, Pusa Snowball KT-25, Dania, Ooty-1

Maturity Groups of Cauliflower

Early I (A) - September Maturity (Mid Sept-Mid Nov)

Nursery Sowing: To ensure a robust and healthy start for cauliflower seedlings, nursery sowing should take place from mid-May to early July. This timing allows the seedlings to establish themselves well before the high temperatures of summer. Early sowing helps in avoiding heat stress and ensures that the plants are ready for transplanting by mid-July, coinciding with the onset of cooler weather which is crucial for cauliflower growth.

Transplanting Time: Transplanting in mid-July allows cauliflower plants to grow during the cooler part of the year, thereby avoiding the detrimental effects of peak summer heat. This period is optimal for plant establishment and growth, ensuring that the plants are well-prepared for curd formation as temperatures begin to drop.

Optimal Temperature Range for Curding: The ideal temperature range for curd formation in cauliflower is between 20-25°C. Maintaining this temperature range helps in preventing heat stress and promotes the development of high-quality curds. This range is critical for achieving the best possible yield and ensuring the marketability of the produce.

Varieties:

- » **Early Kunwari**: Known for its early maturity and good marketability, this variety is favored for its rapid growth and ability to reach the market early.
- » **Pant Gobhi-3**: Offers high yields and quality curds, making it a popular choice for farmers looking to maximize their harvests.
- » **Pusa Meghna**: Suitable for early sowing, this variety is prized for its excellent curd quality and reliable performance.
- » **Pusa Kartik Sankar**: A hybrid variety renowned for its disease resistance and high yield, providing a robust and productive crop.

Early I (B) - October Maturity (Mid Oct-Mid Nov)

Nursery Sowing: Sowing seeds from the end of May to mid-June provides sufficient growing time for seedlings, ensuring they are robust and ready for transplanting by mid-July. This timing allows the plants to establish before the onset of cooler temperatures, which are ideal for curd formation.

Transplanting Time: Transplanting in mid-July ensures optimal growth conditions for the plants, allowing them to develop well before the cooler temperatures of October. This timing helps in achieving a healthy and productive crop.

Optimal Temperature Range for Curding: Maintaining a temperature range of 20-25°C is crucial for high-quality curd formation. This range helps in avoiding heat stress and ensures the production of dense, marketable curds.

Varieties:

- » **Pusa Katki**: Well-suited for mid-season with good yield, this variety is known for its robust growth and productivity.

- **Pusa Deepali**: Recognized for its excellent curd formation, providing high market value due to its quality produce.
- **Pant Gobhi-2**: Offers resistance to common diseases and pests, making it a durable and reliable choice for farmers.

Mid Early (II) - November Maturity (Mid Nov-Mid Dec)

Nursery Sowing: Late July to early September ensures that seedlings are robust and ready for transplanting. This period allows for ample growth before the onset of cooler temperatures, which are ideal for curd formation.

Transplanting Time: Early September transplanting takes advantage of cooler temperatures, which are essential for healthy curd development. This timing ensures that the plants are well-established and ready to produce high-quality curds.

Optimal Temperature Range for Curding: The ideal temperature range for curd formation is 16-20°C. Maintaining this range promotes healthy curd development and ensures the production of high-quality produce.

Varieties:

- **Improved Japanese**: Known for its compact curds and disease resistance, making it a preferred choice for many growers.
- **Pusa Hybrid-2**: A hybrid with high yield potential and good curd quality, offering farmers a dependable option.
- **Pusa Sharad**: Suitable for November maturity with consistent performance, providing reliability in yield.
- **Pant Gobhi-4**: Offers high yield and quality curds, ensuring good returns for farmers.

Mid Late (III) - December Maturity (Mid Dec-Mid Jan)

Nursery Sowing: Late August to late September ensures optimal seedling growth. This period allows seedlings to grow in ideal conditions, ensuring they are robust and ready for transplanting.

Transplanting Time: Early October to mid-November transplanting ensures robust plant development, taking advantage of the cooler weather for optimal growth.

Optimal Temperature Range for Curding: A temperature range of 12-16°C is ideal for preventing bolting and ensuring the formation of compact, high-quality curds.

Varieties:

- **Pusa Synthetic**: Known for its uniform curds and high yield, offering a reliable crop.
- **Pusa Subhra**: Offers resistance to common diseases and pests, making it a durable option for farmers.

» **Palam Uphar**: Suitable for December maturity with consistent performance, ensuring reliable yields.

» **Pant Subhra**: Known for high yield and quality curds, providing good market returns.

» **Pusa HimJyoti**: Offers excellent curd quality and yield, making it a valuable variety for late maturity.

» **Pb Giant 35**: Robust variety with good disease resistance, ensuring healthy crop development.

» **Pusa Paushja**: Known for its uniform curds and high yield, offering a reliable harvest.

» **Pusa Shukti**: Offers resistance to common diseases and pests, providing a durable crop option.

Late (IV) Snowball - January-March Maturity

Nursery Sowing: Late September to mid-October ensures healthy seedlings ready for transplanting during the optimal time frame. This timing allows for robust seedling development in preparation for transplanting.

Transplanting Time: October end to mid-November is the ideal period for transplanting, ensuring that plants are established in time to take advantage of cooler temperatures.

Optimal Temperature Range for Curding: Maintaining a temperature range of 10-16°C is crucial for ensuring compact curds and preventing bolting, resulting in high-quality produce.

Varieties:

» **Snowball 16**: Known for its compact and high-quality curds, making it a popular choice for late-season production.

» **Pusa Snowball-I**: Offers resistance to common diseases and pests, ensuring a healthy and productive crop.

» **Pusa Snowball K-1**: Known for its uniform curds and high yield, providing reliable returns.

» **Pusa Snowball KT-25**: Offers excellent curd quality, making it a preferred choice for high-quality produce.

» **Dania**: Suitable for late-season cultivation with good yield, ensuring reliable production.

» **Ooty-1**: Adapted for cultivation in high-altitude regions with excellent curd quality, making it valuable for specific climatic conditions.

Seedlings

Seedlings typically become ready for transplanting within 4-6 weeks. For successful transplantation, seedlings should have a diameter of 5 mm and a length of approximately

10-12 cm, ensuring they offer better crop stands with minimal mortality rates. Additionally, selecting robust seedlings can significantly impact the overall health and yield of the crop. During this period, it is essential to maintain adequate watering and protection from pests to ensure strong growth. Using high-quality seeds and providing optimal growing conditions can further enhance the seedlings' development. Preparing seedlings in a controlled environment such as a greenhouse can also improve their quality.

Seed Rate

- **Early Varieties:** Require a seed rate of 600-750g per hectare.
- **Mid-Early Season Varieties:** Require 500g per hectare.
- **Mid-Late Varieties:** Require 400g per hectare.
- **Late Varieties:** Require 300g per hectare.

The seed rate is crucial for ensuring optimal plant population and yield. For early varieties, the higher seed rate accounts for faster germination and growth, essential for early market supply. Mid-early season varieties need slightly less seed, balancing between growth speed and plant density. Mid-late varieties require even less, focusing on gradual development over a more extended period. Late varieties need the least seed rate, aligning with their longer growth cycle. Adjusting the seed rate according to variety ensures efficient use of seeds and resources.

Soil Preparation and Transplanting

Preparing the soil involves thorough ploughing, starting with a soil-turning plough followed by 4 to 5 ploughings with a country plough. This process should be complemented with leveling to achieve a fine tilth. Manure application during field preparation is essential for enriching the soil. Effective drainage is particularly important for early and mid-season crops, as rainfall can coincide with these periods. Therefore, early crops should be planted on ridges or raised beds to prevent waterlogging, whereas mid and late varieties can be planted on flat beds. Transplanting should be done in the late afternoon to minimize heat stress on the young plants. Consistent soil preparation and proper timing can significantly enhance plant establishment and growth.

Spacing

- **Early Varieties:** 45 cm x 30 cm.
- **Mid and Late Season Varieties:** 60 cm x 45 cm.

Proper spacing ensures each plant has adequate nutrients, light, and space to grow, reducing competition and promoting healthy development. Early varieties benefit from closer spacing due to their faster growth and shorter maturation period. Mid and late-season varieties require wider spacing to accommodate their larger size and longer growth period. This spacing strategy helps in preventing diseases and pests by improving air circulation and reducing humidity around the plants. Adequate spacing also facilitates easier management practices like weeding, irrigation, and harvesting.

Popular Varieties and Hybrids

Early Varieties

1. **Early Kunwari**
    - **Description:** This variety is well-suited for regions like Punjab, Haryana, Himachal Pradesh, and Delhi. The leaves are bluish-green with a waxy bloom, and the curds are semi-spherical, ready for harvesting from mid-September to mid-October.
    - **Characteristics:** Early Kunwari is recommended for sowing from mid to late May. It has a short growing season, making it ideal for early market supply. Its robust nature ensures good yields even in less-than-ideal growing conditions. The variety is appreciated for its resilience and early maturity, providing a competitive advantage in the market.
2. **Pusa Early Synthetic**
    - **Description:** Developed by IARI, New Delhi, this variety features erect plants with bluish-green leaves. The curds are small to medium-sized, flat, creamy white, and compact. It is suitable for early cultivation in both northern and southern states.
    - **Characteristics:** Pusa Early Synthetic has a high yield potential and shows resistance to riceyness, a common physiological disorder in cauliflower. Its compact growth makes it ideal for dense planting. This variety's adaptability to different climatic conditions ensures consistent yields across various regions. Its early maturity and high quality make it a preferred choice for many farmers.

Mid-Season Varieties

1. **Pusa Deepal**i
    - **Description:** Developed through inbreeding at IARI, New Delhi, Pusa Deepali has compact, self-blanching, white, and medium-sized curds. It is suitable for sowing from late May to early June.
    - **Characteristics:** This variety produces uniform curds that are well protected by leaves, maturing in about 80-90 days. It is particularly suitable for cultivation in northern India, especially in Delhi and Punjab. Pusa Deepali is known for its excellent curd quality and resistance to common pests. Its medium-sized curds are highly marketable, making it a favorite among growers.

2. **Pusa Hybrid-2**
   - **Description:** Recommended by IARI, New Delhi, this variety features semi-erect plants with bluish-green leaves. The curds are creamy-white and highly compact, maturing 120 days after transplanting.
   - **Characteristics:** Pusa Hybrid-2 is resistant to downy mildew and offers a high yield potential of 230-250 q/ha. It is recommended for humid regions like the Bengal-Assam basin and the Sutlej-Ganga alluvial plains. This variety's resistance to diseases and high yield make it a reliable choice for farmers. Its adaptability to different environmental conditions ensures stable production.

Late Varieties

1. **Pusa Snowball-1**
   - **Description:** Suitable for cool seasons, this variety was developed by IARI, Regional Station, Katrain. The leaves are straight and upright, tightly covering the curd. The curds are very compact, medium-sized, and snow white.
   - **Characteristics:** Pusa Snowball-1 is a late variety suitable for sowing from mid-September to the end of October in north India. The optimum temperature for curd initiation and development is 10-16°C. This variety is appreciated for its excellent curd quality and high market value. Its ability to thrive in cooler temperatures makes it ideal for late-season cultivation.
2. **Pusa Snowball K-1**
   - **Description:** Developed at IARI, Regional Station, Katrain, this variety is tolerant to black rot. The leaves are puckered, serrated, and light green. The curds are snow white and retain their color even if harvesting is delayed.
   - **Characteristics:** Pusa Snowball K-1 is slightly later than Pusa Snowball-1, extending the availability period of cauliflower. It has excellent curd quality, making it popular among farmers and consumers. This variety's tolerance to black rot and its extended harvesting window provide flexibility for growers. Its high-quality curds are in demand in the market, ensuring good returns for farmers.

Nutrient Management

For optimal cauliflower growth, the application of manure and fertilizers should be tailored based on soil fertility tests. Typically, 150-200 quintals per hectare of farmyard manure is applied before transplanting. The requirements for nitrogen, phosphorus, and potassium vary by region, with recommended doses of 40-60 kg N/ha, 60-120 kg P2O5/ha, and 60 kg K2O/ha. Micronutrients such as boron and molybdenum are also crucial for preventing disorders like browning and whiptail. Consistent nutrient management not only promotes vigorous growth but also enhances the crop's resistance to diseases and pests. It's important to monitor soil health regularly and adjust the fertilizer application accordingly. By incorporating organic matter and maintaining balanced nutrient levels, farmers can ensure robust and healthy cauliflower crops.

Irrigation

Cauliflower requires consistent moisture to achieve optimal growth. The frequency of irrigation depends on the soil type and climatic conditions. For early and mid-season crops, fewer irrigations may be necessary due to natural rainfall. However, regular irrigation is critical during curd formation to ensure high-quality curds. Maintaining soil moisture levels is essential for preventing stress on the plants, which can lead to poor development and reduced yield. Implementing efficient irrigation practices, such as drip irrigation, can help in conserving water while providing the necessary moisture to the plants. Timely irrigation during critical growth stages can significantly enhance the quality and quantity of the harvest.

Weed Management

Weeds compete with cauliflower for nutrients, light, and water, making regular weeding and hoeing essential for maintaining crop health. Herbicides like Basalin (2-3 kg a.i/ha) and Tenoran (1 kg a.i/ha) can be used for effective weed control. Intercultural operations should be carried out carefully to avoid damaging the roots of the cauliflower plants. Effective weed management ensures that the cauliflower plants receive the maximum nutrients and resources available, promoting healthy growth and higher yields. Additionally, adopting an integrated weed management approach that combines mechanical, cultural, and chemical methods can provide long-term control of weed populations. Regular monitoring and timely interventions are key to keeping the field weed-free.

Blanching

Blanching is necessary to protect cauliflower curds from direct sunlight, which can cause yellowing. This is especially important for early and mid-season varieties with open plant types. Blanching can be achieved by tying the leaves over the curd or placing a leaf over the curd 4-5 days before harvesting. Proper blanching not only preserves the white color of the curds but also improves their market value. Ensuring that the curds are well-covered during their development helps in maintaining their aesthetic quality. Blanching techniques may vary depending on the variety and growing conditions, but the goal remains the same: to produce high-quality, visually appealing curds.

Harvesting, Yield, and Storage

Harvesting should be done when the curds are fully developed and compact. Delayed harvesting can lead to over-maturity, resulting in loose, leafy, or ricey curds. Early maturing varieties yield 80-120 q/ha, while main season and late varieties can yield up to 300 q/ha. Proper post-harvest handling and storage conditions are crucial to maintaining the quality and shelf-life of the curds. Cauliflower can be stored at 0°C and 90-95% relative humidity for 2-4 weeks, ensuring a consistent supply to the market. Maintaining optimal storage conditions helps in preserving the nutritional quality and extending the marketability of the crop. Efficient harvesting and storage practices are vital for maximizing profitability and reducing post-harvest losses.

Diseases, Pests, and Physiological Disorders

Diseases

1. **Stalk Rot (*Sclerotinia sclerotiorum*)**
   - **Symptoms:** Light green leaves turning yellow, black soft rot on petioles and mid-ribs.
   - **Control:** Implement crop rotation, and apply fungicides like benlate (0.02%) and MBC (0.02%). Proper field hygiene and timely removal of infected plant parts can also help in managing the disease. Regular monitoring and early detection are crucial for effective control.
2. **Black Rot (*Xanthomonas campestris* pv. *campestris*)**
   - **Symptoms:** Yellow V-shaped lesions on leaf margins, blackened veins.
   - **Control:** Treat seeds with Plantomycin (100 ppm), Agrimycin 100 (100 ppm), or Streptocyclin (100 ppm). Ensuring good drainage and avoiding overhead irrigation can reduce the incidence of black rot. Using resistant varieties can also provide effective control.

Pests

1. **Diamond Back Moth (*Plutella xylostella*)**
   - **Symptoms:** Slender pale green caterpillars feeding on leaves.
   - **Control:** Foliar spray with Cypermethrin (0.2 kg a.i./ha), Femvalerate (0.1 kg a.i./ha), or Prothiophos (0.75 kg a.i./ha). Biological control methods, such as introducing natural predators, can also be effective. Regular scouting and timely application of control measures are important for managing infestations.
2. **Cabbage Butterfly (*Pieris brassicae*)**
   - **Symptoms:** Caterpillars feeding on young rootlets and main roots.
   - **Control:** Apply sprays of 0.15% Carbaryl and 0.4% fungicide. Maintaining field hygiene and using physical barriers can help in reducing pest populations. Integrated pest management practices can provide sustainable control.

Physiological Disorders

1. **Riceyness**
   - **Symptoms:** Loose, granular curds with elongated peduncles.
   - **Causes:** High or low temperatures, delayed harvesting, heavy nitrogen application.

- **Control:** Grow appropriate varieties and follow proper cultural practices. Monitoring and maintaining optimal temperature conditions during curd development can help in preventing riceyness.

2. **Fuzziness**
   - **Symptoms:** Velvety curds with elongated flower pedicels.
   - **Causes:** Genetic factors, out-of-season cultivation.
   - **Cont**rol: Use genetically pure seed and cultivate in the right season. Adhering to recommended planting schedules can minimize the occurrence of fuzziness.
3. **Leafiness**
   - **Symptoms:** Formation of small leaves from the curd.
   - **Causes:** High temperatures during curding, genetic factors.
   - **Control:** Select suitable varieties and follow proper cultural practices. Maintaining optimal growing conditions can help in reducing leafiness.
4. **Browning (Brown Rot or Red Rot)**
   - **Symptoms:** Water-soaked, light brown to dark brown spots on curds.
   - **Causes:** Boron deficiency.
   - **Control:** Apply borax to the soil (20 kg/ha) and use foliar sprays of borax (0.25-0.50%). Regular soil testing and nutrient management can prevent boron deficiency.
5. **Whiptail**
   - **Symptoms:** Chlorotic, cupped leaves with withered edges.
   - **Causes:** Molybdenum deficiency in acidic soils.
   - **Control:** Apply lime to raise soil pH and use sodium or ammonium molybdate (1-2 kg/ha). Correcting soil pH and ensuring adequate molybdenum levels are essential.
6. **Buttoning**
   - **Symptoms:** Development of small curds with inadequate foliage.
   - **Causes:** Over-aged seedlings, unfavorable weather, poor nursery management.
   - **Control:** Ensure timely transplanting, adequate nitrogen and water supply, and proper nursery management. Consistent monitoring and care during the seedling stage are crucial.
7. **Blindness**
   - **Symptoms:** Plants without terminal buds, failure to form curds.

- **Causes:** Low temperatures, damage to terminal buds.
- **Control:** Avoid exposure to low temperatures and handle seedlings carefully. Protecting plants from extreme weather conditions can prevent blindness.

8. **Chlorosis**
   - **Symptoms:** Interveinal chlorosis and yellow mottling of lower leaves.
   - **Causes:** Magnesium deficiency in acidic soils.
   - **Control:** Apply magnesium oxide (300 kg/ha) and use dolomite limestone to raise soil pH. Regular soil amendments can correct magnesium deficiency.
9. **Hollow Stem**
   - **Symptoms:** Hollow stems with water-soaked, discolored tissue.
   - **Causes:** Boron deficiency, high nitrogen levels.
   - **Control:** Apply borax to the soil (15-20 kg/ha) and use foliar sprays of borax (0.1-0.3%). Balanced fertilization practices can prevent hollow stem.
10. **Frost Injury**
   - **Symptoms:** Yellowish-white leaves, flaccid petioles.
   - **Causes:** Exposure to frost.
   - **Control:** Irrigate the field before frost and create smoke to raise the field temperature. Protecting plants from frost can prevent damage.
11. **Pinking**
   - **Symptoms:** Pink tinge on curds.
   - **Causes:** High light intensity exposure.
   - **Control:** Provide proper shading and protection from direct sunlight. Ensuring adequate coverage during peak sunlight hours can prevent pinking.

Cauliflower is a valuable crop with significant nutritional, economic, and health benefits. Proper cultivation practices, including soil preparation, timely sowing and transplanting, nutrient management, and pest and disease control, are essential for achieving high yields and quality curds. Understanding and managing physiological disorders is crucial to ensure a successful cauliflower crop. With its adaptability to diverse climatic conditions, cauliflower continues to be a vital component of global agriculture and cuisine. Proper management practices and attention to detail can help in maximizing productivity and ensuring the sustainability of cauliflower farming.

c. Knol-Khol

Brassica oleracea* var. *gongylodes

Family: Brassicaceae

Introduction

Knol-khol, also known as kohlrabi, is a distinctive member of the Brassica family, botanically referred to as *Brassica oleracea* var. *gongylodes*. It is characterized by its unique swollen stem, forming a knob-like structure above the ground, often called the tuber. This knob is prized for its crisp, juicy texture and mild flavor, making it a versatile vegetable that can be enjoyed both raw and cooked. Knol-khol holds popularity in various regions across India, including Kashmir, West Bengal, Maharashtra, Assam, Uttar Pradesh, Punjab, and parts of South India. Despite its popularity, extensive commercial cultivation is limited.

Knol-khol is valued for its nutritional benefits, being rich in vitamins and minerals, particularly vitamin C and dietary fiber. Its versatility in culinary applications, from salads to stews, makes it a staple in many diets. Additionally, its relatively easy cultivation process allows even small-scale farmers to grow this crop effectively. The vegetable's adaptability to different climatic conditions further enhances its potential for diverse agricultural practices.

Climate and Soil

Knol-khol thrives best in a relatively cool, moist climate, making it ideal for temperate regions but also suitable for subtropical areas. The optimal temperature range for its cultivation is between 15°C and 20°C, with an average maximum temperature of 24°C and a minimum of 4.5°C. Early varieties are sensitive to bolting, particularly in temperate conditions where low night temperatures can induce premature flowering. However, in subtropical regions, the diurnal temperature variation helps mitigate this effect, reducing the risk of bolting.

Knol-khol exhibits a remarkable tolerance to extreme cold and frost, outperforming many other cool-season crops in such conditions. It grows best in soils rich in organic matter, preferring sandy loam for early crops and clay or silt loam for higher yields and late crops. The ideal soil pH for knol-khol ranges between 5.5 and 6.8, and it performs poorly in highly acidic soils. Proper soil preparation, including adequate organic matter incorporation, can significantly enhance crop productivity and quality.

Varieties

Several promising knol-khol varieties are cultivated in India, each with unique characteristics suited to different growing conditions:

1. **Sutton's Earliest Purple**: Known for its early maturity and resistance to bolting. This variety is ideal for regions where early harvesting is crucial to market supply.

2. **Golith White**: Popular for its large, white knobs, offering a visually appealing and high-yield option for farmers.
3. **Early White Vienna**: Features medium-sized, light green knobs with tender, creamy-white flesh, making it a favorite for fresh consumption.
4. **King of the Market**: Recognized for its excellent yield and knob quality, suitable for markets demanding high-quality produce.
5. **Early Purple Vienna**: Produces purplish-blue knobs with greenish-white flesh, providing a unique color contrast that appeals to consumers.

Recommended Varieties

1. **King of the North**
   - **Description**: This variety is characterized by dark green foliage, flattish-round knobs, and a large leaf sheath that spreads over the knob.
   - **Maturity**: 60-65 days after transplanting.
   - **Characteristics**: Known for its robust growth and adaptability to various climatic conditions, making it a reliable choice for different regions. Its strong growth habit ensures good yield stability and quality.
2. **Large Green**
   - **Description:** Produces green, round, large-sized knobs with small tops. The knobs are tender and delicately flavored with white flesh.
   - **Maturity:** 76 days with an average yield potential of 225-250 quintals per hectare.
   - **Recommended For:** Suitable for mid and high hills of the western Himalayas, this variety is known for its adaptability to higher altitudes and cooler climates, making it a preferred choice for these regions.
3. **Purple Vienna**
   - **Description:** Matures slightly later than White Vienna, with purplish-blue knobs and greenish-white flesh.
   - **Maturity:** 55-65 days with a slightly better yield potential than White Vienna.
   - **Characteristics:** Its unique color and early maturity make it a market favorite, offering an attractive option for both fresh markets and processing.
4. **White Vienna**
   - **Description:** An early variety with globular, light green, smooth, tender, medium-sized knobs with creamy-white flesh.
   - **Maturity:** 55-65 days with a yield potential of 150-200 quintals per hectare.

- **Characteristics:** Features dwarf plants with medium green leaves and stems, making them suitable for dense planting. Its early maturity and tender flesh make it a popular choice for fresh consumption.

Knol-khol's diverse varieties and adaptability make it a valuable crop for different regions and climatic conditions. By selecting the appropriate variety and following best cultivation practices, farmers can achieve high yields and quality produce, meeting both market demands and nutritional needs.

Cultivation Practices

Propagation and Seedling Preparation

Knol-khol is typically propagated by seed, with an optimal seed rate of 1-1.5 kg/ha. To prevent black rot, seeds should be treated with hot water at 50°C for half an hour. Additionally, treating seeds with Apron 35 (2 g/kg seed) helps protect against downy mildew. Seedlings are raised in nursery beds, and 4-6 week-old seedlings are ready for transplanting.

» **Nursery Bed Preparation:** Nursery beds should be 60 cm wide and 2.5 m long, with a soil mixture containing 15 g each of nitrogen, phosphorus, and potassium, and 2.5-4 kg of farmyard manure per square meter. For light and drought-sensitive soils, sunken nursery beds are preferred. Acidic soils should be limed. Treating nursery beds with formalin (40% formaldehyde diluted in 5-6 parts water) or Captan (0.3%) for soil drenching helps prevent fungal diseases. Proper bed preparation ensures healthy seedlings, leading to better crop establishment.

» **Sowing:** Seeds are sown in rows 5-6 cm apart at a depth of 1.5-2 cm. The beds are covered with grass or polythene to conserve moisture and ensure uniform germination. Mulch is removed just before germination to control damping off, which can be managed with Dithane M-45 (0.2%). Consistent monitoring and timely interventions can significantly improve germination rates and seedling vigor.

Transplanting

In the plains of North India, transplanting is done in September, while in regions with milder winters, October is the best time. In the hills of Northern India, seeds are sown from March to August. About 5-6 week-old seedlings are transplanted for summer and autumn crops.

» **Land Preparation:** The land is plowed to a fine tilth with 2-3 plowings, followed by the preparation of beds and channels for irrigation. Transplanting is done in the evening or on cloudy days, ensuring the soil around the plant is well pressed to establish contact with the roots. Light irrigation is provided immediately after transplanting. Proper land preparation and transplanting techniques ensure high survival rates and vigorous growth.

» **Spacing:** Transplanting is done at a spacing of 25 cm × 25 cm, 25 cm × 30 cm, 25 cm × 40 cm, or 30 cm × 45 cm, depending on climatic conditions and soil fertility. Closer spacing results in higher yields but smaller knobs. Early varieties are planted closer, while late ones require wider spacing. Proper spacing minimizes competition for nutrients and sunlight, promoting healthy plant development.

Manuring and Fertilization

Knol-khol is a heavy feeder, requiring substantial nutrients for optimal growth. A yield of 20 tonnes/ha removes approximately 100 kg N, 85 kg P, and 170 kg K from the soil. Excess nitrogen can cause abundant leafy growth and delay the crop. Split application of nitrogen is beneficial, with half applied at transplanting and the rest in two equal doses: 3 weeks after transplanting and at the knob development stage.

» **Farmyard Manure:** Apply 150-200 q/ha of farmyard manure 4-6 weeks before transplanting. This enhances soil fertility and improves the soil structure.

» **Nutrient Management:** Ensure balanced application of N, P, and K. Foliar application of urea (1-2%) can correct nitrogen deficiency. Micronutrient deficiencies can be addressed with Multiplex (0.2-0.3%) sprays or soil application of borax (10-15 kg/ha) and ammonium molybdate (200-300 g/ha). Regular soil testing and appropriate nutrient management are crucial for optimal crop performance.

Aftercare

Steady growth is crucial for high-quality knobs. Any check in growth can result in fibrous, woody knobs. Rapid growth after a slow start can cause cracking. Elongated knobs may develop due to close spacing, high temperatures, or excess nitrogen.

» **Intercultural Operations:** Regular hoeing and weeding are essential to maintain soil moisture and aeration. Shallow hoeing helps control weeds without damaging the shallow root system. Once the soil is covered with foliage, hoeing is stopped, and any remaining weeds are removed by hand. Proper intercultural operations enhance soil health and plant growth.

» **Weed Management:** Herbicides like Treflan (Trifluralin @ 0.5 L/ha), Semeron (Desmetrayne @ 1 kg/ha), and Stomp (Pendimethalin @ 1-2 kg/ha) are effective. A combination of herbicide application and manual hoeing is economical and efficient. Integrated weed management practices ensure effective and sustainable weed control.

Irrigation

Knol-khol requires a consistent supply of moisture for uniform growth and knob development. The first irrigation is done immediately after transplanting, followed by subsequent irrigations as needed, depending on soil and weather conditions. Generally, irrigations at 15-day intervals are adequate. Heavy irrigation should be avoided, and moisture content should be maintained below 50% of field capacity at the maturity stage to prevent detrimental effects on the knobs.

Proper irrigation management is crucial for achieving high yields and quality produce.

Harvesting and Post-Harvest Management

Knobs are harvested when they are fully developed but still tender, typically at 5-8 cm in diameter. Overgrown knobs become tough and woody. Harvesting is done by cutting the stem just below the knob with a sharp knife or sickle.

- » **Yield:** Early varieties produce knobs weighing 200-250 g, while late varieties can yield knobs up to 1 kg. Yields vary from 12 to 30 tonnes per hectare. High yields are achieved through proper cultivation practices and timely interventions.
- » **Post-Harvest Handling:** Knobs are marketed either with or without leaves, depending on consumer preference. Proper post-harvest handling and storage are crucial to maintaining the quality and shelf-life of the knobs. Effective post-harvest practices ensure that the produce reaches the market in the best possible condition, maximizing returns for the farmers.

4. ONION

***Allium cepa* L.**

(2n = 2 x = 16)

(Hindi: Pyaz)

Allium cepa L. (Onion)

The onion (*Allium cepa L.*), commonly known in Hindi as Pyaz, is a staple in many kitchens worldwide due to its distinctive odor, flavor, and pungency. This pungency arises from allyl-propyl-disulphide, a volatile oil produced enzymatically when the onion tissues are damaged. Onions are highly valued for their storage longevity and transport resilience, making them a global commodity. They are consumed in various forms: raw in salads, cooked in dishes, or processed into dehydrated onions and flakes. Additionally, onions are nutrient-rich, particularly in phosphorus and calcium, and have numerous medicinal applications, such as diuretic properties and topical uses for wounds.

India ranks as the second-largest producer of onions globally, following China, and holds a significant position in global exports. The major states contributing to its production include Maharashtra, Karnataka, and Gujarat.

Origin and Taxonomy

Allium cepa is believed to have originated in Asia, with historical references suggesting areas including Pakistan and nearby mountainous regions as primary centers of origin. The species is diploid with a chromosome number of 2n = 2x = 16.

Varieties of Onions Cultivated in India:

1. **Common Onion (*Allium cepa var. cepa*)**: Large bulbs, propagated by seeds, mainly used raw or cooked.
2. **Multiplier Onion or Potato Onion (*Allium cepa var. aggregatum*)**: Small bulbs in clusters, propagated through bulbs, used for seasoning.
3. **Shallot (*Allium cepa var. ascalonicum*)**: Clustered bulbs, rarely produces seeds, propagated through bulbs.
4. **Tree Onion or Egyptian Onion (*Allium cepa var. viviparaum/proliferum*)**: Perennial, produces bulb-lets instead of seeds, used for pickling.
5. **Chive (*A. schoenoprasum*)**: Perennial herb, grown for its leaves, propagated by root division.

Botany of the Onion

Onion plants are biennial with tubular leaves. The bulb, which forms the main edible part, is an aggregation of swollen leaf bases attached to a rudimentary underground stem. The flower structure, or umbel, comprises multiple inflorescences and is crucial for seed production. Onions exhibit protandry, requiring cross-pollination predominantly facilitated by bees.

Onion Valuation and Chemical Properties: Onions are highly valued globally for their distinctive characteristics—odor, flavor, and pungency—which make them a staple in culinary uses across various cultures. The pungency of onions is primarily due to allyl-propyl-disulphide, a volatile oil that is released when the onion's tissues are disrupted. This compound is part of a group of sulfur-containing compounds that are responsible for the tears and intense flavors associated with cutting onions. The enzymatic processes that release these compounds are activated upon slicing the onion, which then combines with air to intensify the pungent aroma and taste.

Storage and Transport Capabilities: One of the practical virtues of onions is their suitability for long-term storage and resilience during long-distance transport. Their layers of dry skin provide a natural barrier to moisture loss and microbial invasion, which contributes to their longevity. This makes onions particularly valuable in areas where food preservation without refrigeration is crucial. Moreover, their robust nature allows them to withstand the rigors of transportation over vast distances without significant quality degradation, making them an excellent commodity for international trade.

Culinary Uses: Onions are incredibly versatile in culinary applications. They can be consumed raw, adding crunch and zest to salads, or they can be cooked—fried, boiled, baked, or caramelized—to enhance their sweetness and add depth to the flavor profile of dishes. Onions are essential in the base flavors of many cuisines, used in everything from European stews and Indian curries to American burgers and Chinese stir-fries. Furthermore, onions are processed into forms such as dehydrated onions and onion flakes, which are used in seasoning blends, instant meals, and snacks, providing flavor and convenience.

Nutritional and Medicinal Benefits: Onions not only enrich the flavor of dishes but also offer various health benefits. They are a good source of vitamin C, dietary fiber, and folic acid. They also contain calcium and phosphorus, which are important for healthy bones and teeth, as well as compounds that are believed to have anti-inflammatory, anti-cholesterol, antioxidant, and anticancer properties. Medicinally, onions have been used historically to treat ailments like headaches, heart disease, and mouth sores, and they possess diuretic properties that help increase urine output.

Global Production and Economic Impact: India is a major player in the global onion market, second only to China in terms of production. The extensive cultivation across Indian states like Maharashtra, Karnataka, and Gujarat is driven by both domestic consumption and significant demand from international markets. The adaptability of onions to diverse climatic conditions allows for widespread cultivation. India's prominence in the onion export market, ranking third globally, highlights its economic impact, contributing substantially to the agricultural export revenues.

Male Sterility in Onions

Male sterility in onions results from the interaction between a nuclear gene and a cytoplasmic factor. Sterile plants, useful in hybrid seed production, carry 'S' cytoplasm and are typically maintained through specific breeding lines known as 'A', 'B', and 'C' lines, which help preserve or restore fertility.

Climatic Requirements

Onions thrive in mild climates without extreme temperatures or excessive rainfall. They are long-day plants, requiring specific photoperiods for bulb development, though varieties vary in their day-length requirements. Optimal growing temperatures range from 12.8°C to 25°C.

Cultivation Methods

Onions can be planted using various methods depending on local conditions:

- **Transplanting Seedlings**: Common for high-yield cultivation, involves raising seedlings in nurseries before transplanting.
- **Direct Bulb Planting**: Used in regions with slopes or terraces, utilizes medium-sized bulbs.
- **Broadcasting or Drilling Seeds**: Employed to save labor, requires careful weeding and thinning.

Nutrient Management

Onions are heavy feeders requiring balanced nutrition, particularly of nitrogen and potassium, for optimal growth and yield.

Harvesting and Post-Harvest Handling

Harvest times vary with the onion variety and planting time. Post-harvest, onions should be cured properly to enhance storage life and minimize losses due to sprouting or rot. Effective storage methods and periodic maintenance like turning the bulbs are crucial for prolonging freshness and preventing decay.

Economic Importance

Onions are a significant agricultural product in India, not only fulfilling domestic needs but also serving a substantial export market. The export market is diversified, catering to preferences for size, color, and pungency in different regions.

Varietal Table for Common Onions

Variety	Origin	Characteristics	Yield (t/ha)
Arka Niketan	Bangalore	Pink, globular bulbs, highly pungent	42
Pusa Red	New Delhi	Red, less pungent, good storage	25-30
N-2-4-1	Rahuri	Brick red, suitable for rabi season	30

Improved Varieties of Onions

Developing Institution	Variety	Special Features	Yield (t/ha)	Harvest Duration (days)
IIHR, Bangalore	Arka Pragathi	Bulbs are globular, deep pink, highly pungent, good storage quality.	45	130
IIHR, Bangalore	Arka Niketan (Sel-13)	Globular, pink bulbs, thin neck, very pungent, excellent storage quality.	42	145
IIHR, Bangalore	Arka Kalyan (Sel-14)	Globular, pink bulbs, suitable for kharif season, moderately resistant to purple blotch.	47	140
IARI, New Delhi	Pusa Red	Red, globular bulbs, less pungent, excellent storage quality.	25-30	125-140
IARI, New Delhi	Pusa White Flat	Ideal for dehydration and green onion, medium to large bulbs.	33-35	125-140
MPKV, Rahuri	N-2-4-1	Brick red, globular, pungent, firm, good storage quality.	30	140
Punjab Agricultural University, Ludhiana	Punjab Naroya	Red, medium to large, round, thin neck, tolerant to purple blotch.	37.5	123

NHRDF, Nasik	Agrifound Dark Red	Dark red, globular, moderately pungent, suitable for export, excellent keeping quality.	30-40	160-165
HAU, Hisar	Hisar-2	Light red bulbs, suited for various culinary uses.	20	-
IARI, New Delhi	Pusa Ratnar	Bronze deep red, flat globular, less pungent, excellent storage quality.	32.5-35	145-150
Punjab Agricultural University, Ludhiana	Punjab Selection	Red, globular, medium size, excellent keeping quality.	30	-
IIHR, Bangalore	Arka Bindu (Rose onion)	Small, crimson red bulbs, 12.5% TSS, good for early harvest and market.	25	90-95

Hybrid Varieties Developed by IIHR, Bangalore

Hybrid Name	Description	Yield (t/ha)	Suitable Seasons	Export Potential
Arka Kirtiman	F1 hybrid, medium size, dark red bulbs, good keeping quality.	45-60	Kharif and Rabi	Suitable for export to Gulf countries
Arka Lalima	F1 hybrid, globe-shaped, deep red, good shelf life of 5 months.	40-50	-	Suitable for export to Gulf countries
Arka Pitamber	Short day variety, globe-shaped yellow bulbs, moderately resistant to purple blotch.	35-38	-	-

Transplanting Method Detailed Explanation

1. Seed Sowing

The transplantation process begins with the careful sowing of seeds in nursery beds. These nursery beds are meticulously prepared to dimensions of 90-120 cm in width and 7.5-10.0 cm in height, with lengths adjusted to suit the field's size and layout. The timing of seed sowing varies geographically: in the plains, it typically occurs around October-November for the Rabi crop, while in hilly regions, it happens from March to June. The seed rate generally ranges from 8 to 10 kg per hectare, depending on the onion variety and the specific conditions of the nursery. This method ensures a high germination rate and healthy seedling development.

» **Soil Preparation:** The soil in the nursery beds is enriched with organic matter and nutrients to support seedling growth. Adding well-decomposed farmyard manure or compost enhances soil fertility.

» **Sowing Technique:** Seeds are sown in rows 5-6 cm apart at a depth of 1.5-2 cm.

Light covering with soil or sand helps retain moisture and provides a conducive environment for germination.

» **Moisture Management:** The beds are watered regularly to maintain consistent moisture levels, essential for uniform germination and seedling growth.

2. Nursery Management

Seedlings are nurtured in these nurseries under controlled conditions to ensure robust growth. Ideal seedlings for transplanting are those that reach about 15 cm in height with a neck diameter of approximately 0.8 cm. This growth stage is usually achieved within 6 to 10 weeks, influenced by factors such as soil quality, climate, and rainfall. Overgrown seedlings may be topped—trimmed at the top to encourage a more robust growth form—just before transplanting to ensure uniformity and manageability.

» **Environmental Control:** Managing temperature and humidity levels in the nursery is crucial for optimal seedling development. Shade nets or polycovers can be used to protect seedlings from extreme weather conditions.

» **Fertilization:** Balanced fertilization, including micronutrient supplements, is provided to promote healthy seedling growth. Foliar sprays of urea or micronutrient solutions can be applied as needed.

» **Pest and Disease Management:** Regular monitoring and preventive treatments help protect seedlings from pests and diseases. Organic or chemical treatments can be used based on the severity of the problem.

3. Land Preparation

Concurrently, the main field is prepared by deep plowing, leveling, and breaking up soil clods to create a fine tilth. This preparation is crucial for facilitating root penetration and moisture retention. The field is then divided into plots of convenient sizes, tailored to optimize irrigation and management practices.

» **Soil Enrichment:** Incorporating organic matter such as compost or well-rotted manure improves soil structure and fertility.

» **Plowing and Leveling:** The field is plowed to a depth of 15-20 cm to ensure good root penetration. Leveling the field helps in uniform water distribution.

» **Plot Layout:** Dividing the field into manageable plots facilitates efficient irrigation and reduces soil erosion. Raised beds or ridges can be prepared in areas prone to waterlogging.

4. Transplanting

The seedlings are carefully transplanted at a spacing of 15 x 8-10 cm. This spacing is critical to allow adequate room for each plant to develop without competition for nutrients, light, and air. Transplanting is done in the evening or on cloudy days to minimize transplant shock and ensure better establishment.

- **Handling Seedlings:** Seedlings are gently uprooted from the nursery bed, retaining as much root soil as possible to reduce transplant shock.
- **Transplanting Technique:** Seedlings are planted at the specified spacing, ensuring the soil around the roots is well-pressed to establish good contact. Light irrigation is provided immediately after transplanting to settle the soil and promote root establishment.
- **Irrigation Management:** Initial irrigation is critical to help seedlings recover from transplant shock. Subsequent irrigations are scheduled based on soil moisture conditions and weather forecasts.

Additional Planting Methods

1. Planting of Bulbs

In regions prone to erosion or where seedlings might be washed away by rain, such as hill slopes or terraced fields, planting medium to small-sized bulbs directly into the field is preferred. This method typically occurs in September-October after the bulbs have rested for a month post-harvest. Bulbs are planted at 15 cm apart in rows on ridges or beds, depending on the terrain and soil type.

- **Site Preparation:** Ridges or raised beds are prepared to prevent waterlogging and soil erosion.
- **Bulb Selection:** Medium to small-sized bulbs are chosen for planting to ensure uniform growth and development.
- **Planting Technique:** Bulbs are planted at the recommended spacing, ensuring they are well-covered with soil to promote root development.

2. Broadcasting or Drilling

This method involves directly sowing seeds into the field by broadcasting or drilling, spaced about 30 cm apart. It is a labor-saving technique but requires vigilant weed management, especially in the early growth stages, to prevent competition.

- **Seed Distribution:** Seeds are evenly distributed over the field using broadcasting or drilling equipment.
- **Soil Coverage:** Light raking or harrowing is done to cover the seeds with a thin layer of soil, ensuring good soil-seed contact.
- **Weed Management:** Pre-plant herbicide application and early-stage manual weeding are crucial to control weed competition and promote healthy crop growth.

Nutrient Management

Onions are heavy feeders, requiring substantial amounts of nitrogen, phosphorus, potassium, magnesium, and sulfur. The nutrient management strategy typically involves:

» **Base Application:** Applying 20-25 tonnes of farmyard manure during initial plowing, followed by the entire doses of phosphorus and potassium. This enriches the soil with essential nutrients and organic matter.

» **Top Dressing:** Nitrogen is applied in two splits, the first 3-4 weeks post-transplanting and the second two months later, to support vigorous growth and bulb development. Foliar applications of urea (1-2%) can correct nitrogen deficiencies.

Irrigation and Weed Control

Irrigation frequency is adjusted based on the crop's growth stage and climatic conditions, with intervals ranging from 13-15 days in the early stages to 7-10 days as the crop matures. Weed control is vital in the early stages, often involving pre-plant herbicide application and manual weeding to ensure that the onions are not out-competed.

» **Irrigation Management:** Regular irrigation schedules are maintained to ensure consistent soil moisture, avoiding both water stress and waterlogging.

» **Weed Control:** Early-stage weed management is crucial for reducing competition. Pre-plant herbicide applications and regular manual weeding are effective strategies.

Harvesting

Onion bulbs are generally ready for harvest 3-5 months after transplanting. The timing of harvest can vary with the variety and seasonal conditions. Harvesting is done manually by pulling bulbs when the tops begin to droop but are still green.

» **Harvest Timing:** Harvesting is timed to ensure bulbs are fully mature but not overgrown, which can affect quality.

» **Harvest Technique:** Bulbs are carefully pulled or dug out to minimize damage and ensure high-quality produce.

Post-Harvest Handling

After harvest, onions undergo curing to dry out the necks and outer layers, enhancing their storage life and market quality. This process is critical for reducing losses due to rotting and sprouting during storage.

» **Curing Process:** Onions are cured by placing them in well-ventilated areas for several days to allow the outer layers to dry and harden.

» **Storage Conditions:** Properly cured onions are stored in well-ventilated conditions to maintain their quality and extend shelf life. For long-term storage, onions can be processed further, such as through dehydration.

These cultivation practices, from seed sowing to post-harvest handling, ensure that onions are grown efficiently and produce high-quality yields. By following these methods, farmers can achieve successful and sustainable onion production.

5. Garlic (*Allium sativum L.*)

Botanical Name: *Allium sativum L.*

Family: Amaryllidaceae

Origin: Central and South Western Asia

Historical and Geographic Origin

Garlic is believed to have originated in the mountainous regions of Central Asia, spanning from southern Turkmenistan northeastward to the Pamir-Alai and Tien Shan regions. It spread during prehistoric times to the Mediterranean region, where it remains highly valued for its culinary and medicinal properties. The wild ancestor of cultivated garlic is thought to be *Allium longicuspis* Rgl. Over centuries, garlic has been integrated into various cultures and cuisines, renowned for its flavor and health benefits. Historical records show that garlic was used by ancient civilizations such as the Egyptians, Greeks, and Romans, underscoring its longstanding significance.

Soil and Climate Requirements

Soil:

- Garlic requires soil that is richer and more fertile than what is typically needed for onions.
- It thrives best in well-drained loamy soils, which provide the necessary nutrients and structure for optimal growth.
- The ideal soil pH for garlic cultivation ranges between 6 and 7, ensuring proper nutrient availability and uptake.
- Garlic does not perform well in highly alkaline or saline soils, which can impede growth and reduce yields.
- Regular soil testing and amendments can help maintain the ideal soil conditions for garlic cultivation.
- Incorporating organic matter, such as compost or well-rotted manure, can enhance soil fertility and structure, supporting robust garlic growth.

Climate:

- Garlic is frost-hardy and requires a cool, moist period during its growth phase, followed by a dry period for bulb maturity.
- Suitable varieties exist for both long-day conditions (around 30°N latitude) and

short-day conditions, allowing for flexibility in cultivation across different regions.

» Cool growing periods generally yield better results compared to warm periods, as they support healthy leaf and root development.

» The critical day length for bulbing in garlic is approximately 12 hours. This photoperiod sensitivity plays a key role in bulb formation.

» Exposing dormant cloves or young plants to temperatures around 20°C or lower for 1-2 months can hasten the bulbing process, enhancing yield and quality.

» Managing the growing environment to ensure these conditions can significantly improve garlic production.

» Utilizing mulching techniques can help retain soil moisture during cool periods and protect the plants from extreme temperature fluctuations.

Improved Varieties of Garlic

Variety	Developing Institution	Characteristics	Yield (tonnes/ha)	Region Recommended
Agrifound White (G 41)	NHRDF	Compact bulbs, silvery-white skin, creamy flesh, 20-25 cloves/bulb, TSS 41%.	13	Areas with less purple blotch or stemphylium blight
Yamuna Safed (G 1)	NHRDF	Compact bulbs, silvery-white skin, 25-30 sickle-shaped cloves, TSS 38%.	15-17.5	All over India
Yamuna Safed 2 (G 50)	NHRDF	Compact bulbs, white-creamy flesh, 35-40 cloves/bulb, TSS 38-40%.	15-20	Northern India
G 282	NHRDF	Large creamy-white bulbs (5-6 cm diameter), 15-16 cloves/bulb, TSS 38-42%.	17.5-20	Northern and Central India
Agrifound Parvati (G 313)	NHRDF	Large bulbs (5-6 cm diameter), creamy-white with pinkish tinge, 10-16 cloves/bulb.	17.5-22.5	Hills of Northern India

Planting Techniques

Sowing Time by Region

Region	Sowing Time
North India	September - November
Maharashtra, Karnataka, Andhra Pradesh	August - November
West Bengal, Orissa, Gujarat	October - November

Sowing times are chosen to ensure that garlic plants have optimal growing conditions. In North India, the September to November window allows the plants to take advantage of the cooler weather. In Maharashtra, Karnataka, and Andhra Pradesh, sowing from August to November ensures the plants benefit from the monsoon rains initially and then the cooler weather as they mature. Similarly, in West Bengal, Orissa, and Gujarat, the October to November sowing time aligns with post-monsoon cooler conditions, which are favorable for garlic growth and development.

Transplanting Techniques

1. **Dibbling:** Cloves are planted 5-7.5 cm deep with the growing ends upwards. This method ensures the cloves are correctly oriented, leading to better sprouting and growth. It is a precise planting method that minimizes damage to the cloves and ensures uniform depth, which helps in achieving consistent growth rates across the field.
2. **Furrow Planting:** Cloves are dropped into furrows by hand and covered lightly with soil. This technique allows for efficient planting in larger fields and ensures that each clove has enough soil contact for good root development. Furrow planting is especially useful in areas with heavier soils, as it facilitates better drainage and aeration around the roots.

Planting Distance

- **Rows:** 15-20 cm apart
- **Plants:** 10 cm apart
- **Sowing Depth:** 2-4 cm

Proper spacing is crucial to avoid overcrowding, which can lead to competition for nutrients and water. By maintaining the recommended planting distances, each garlic plant has enough space to develop a robust root system and bulb. Additionally, adequate spacing helps in air circulation, reducing the risk of fungal infections and promoting healthier plant growth. Closer spacing might be employed in fertile soils with intensive nutrient management to maximize yield per hectare.

Nutrient Management

Manures and Fertilizers

- **NPK Requirements:** 60-125 kg N, 35-65 kg P, and 0-100 kg K per hectare.
- **Organic Manure:** Apply 10-50 tonnes/ha of farmyard manure mixed into the soil during preparation. This enriches the soil with organic matter, enhancing its structure and fertility. Organic manure also improves the soil's water-holding capacity, which is beneficial during dry periods.

- **Micro-Nutrients:** Applications of $MnSO_4$ (0.1%), boric acid (0.2%), $CuSO_4$ (0.02%), or $ZnSO_4$ (0.02%) stimulate dry matter accumulation in the cloves. Borax application up to 10 kg/ha can increase bulb size and yield significantly. Regular monitoring and soil tests can help in adjusting the micronutrient applications to ensure optimal growth conditions.

Irrigation and Weed Management

Irrigation

- Needs irrigation at 8-day intervals during vegetative growth and 10-15 day intervals during maturation.
- Irrigation should be stopped as the crop matures to prevent root and bulb scale rot. Maintaining a consistent irrigation schedule ensures that garlic plants receive adequate moisture without being waterlogged, which is crucial for healthy growth and bulb formation. During the maturation phase, reducing irrigation helps in hardening the bulbs and reducing the risk of diseases associated with high soil moisture levels.

Weed Management

- **First Weeding:** One month after planting.
- **Second Weeding:** Two months after planting.
- **Recommended Herbicide:** Pendimethalin @ 3.35 litres/ha after planting and before the first irrigation, plus one hand weeding at 45 days after planting.

Effective weed management is essential for ensuring that garlic plants do not face competition for nutrients and water. Early weeding helps in controlling weed growth during the critical stages of garlic development. Using a combination of herbicides and manual weeding ensures that the field remains weed-free, which is crucial for maximizing yield and quality.

Harvesting and Post-Harvest Management

Harvesting

- Garlic is ready for harvesting when tops turn yellow or brown and show signs of drying up, typically 4-5 months after planting.
- Manual harvesting is common in India. Ensuring that garlic is harvested at the right time prevents over-maturity, which can lead to splitting and reduced storage life. Proper harvesting techniques help in maintaining the quality of the bulbs, which is crucial for market acceptance.

Yield

- **Yield:** 100-150 quintals per hectare. Achieving high yields depends on following best practices in cultivation, including proper nutrient management, irrigation, and pest control. Consistent monitoring and timely interventions can help in achieving optimal yields.

Post-Harvest Handling

- **Curing:** Sorting, grading, and curing are essential to minimize post-harvest losses. Proper drying and curing improve storage life and quality.
- **Storage:** Store in well-ventilated rooms using nylon-netted bags to minimize losses. Ideal storage conditions are 0.5°C with 60-70% relative humidity.

Proper post-harvest handling is crucial to maintaining the quality of garlic and extending its shelf life. Curing helps in reducing the moisture content of the bulbs, which prevents rotting and sprouting during storage. Ensuring good ventilation and using appropriate storage materials can significantly reduce post-harvest losses.

Plant Protection

Diseases and Management

1. **Purple Blotch (*Alternaria porri*):**
   - **Symptoms:** Whitish sunken lesions with purple centers on leaves.
   - **Control:** Spray Mancozeb @ 2.5 g/litre of water at 15-day intervals. Regular monitoring and timely spraying can help in managing the disease effectively, ensuring healthy leaf growth and better yields.
2. **Stemphylium Blight (*Stemphylium vesicarium*):**
   - **Symptoms:** Yellow to orange flecks or streaks on leaves.
   - **Control:** Spray recommended fungicides at regular intervals. Preventive measures and maintaining field hygiene can also reduce the incidence of this blight.
3. **Cercospora Leaf Blight (*Cercospora duddiae*):**
   - **Symptoms:** Ash-colored spots on leaf lamina.
   - **Control:** Spray Ziram or Captan @ 2.0 g/litre of water. Regular applications and good cultural practices can help in controlling this disease.
4. **Powdery Mildew (*Leveillula taurica*):**
   - **Symptoms:** Pale-yellow patches on the abaxial leaf surface.
   - **Control:** Spray sulfur fungicides @ 2.0 g/litre of water. Early detection and consistent treatment can prevent the spread of powdery mildew.
5. **Mosaic Disease:**
   - **Symptoms:** Chlorotic mottling and strips on leaves.

- **Control:** Spray Monocrotophos @ 0.5 ml/litre of water. Managing vector populations and using virus-free planting material can help in controlling mosaic disease.

Insect Pests and Management

1. **Thrips (*Thrips tabaci*):**
    - **Symptoms:** Leaf surfaces turn white or silver.
    - **Control:** Spray Malathion @ 1 ml/litre or Methyldemeton @ 0.75 ml/litre of water. Regular monitoring and integrated pest management practices can effectively control thrip populations.
2. **Stem and Bulb Nematode (*Ditylenchus dipsaci*):**
    - **Symptoms:** Basal swelling and spongy tissue.
    - **Control:** Pre-plant soil fumigation with dichloropropene-dichloropropane mixture @ 500 kg/ha. Crop rotation and resistant varieties can also help in managing nematode infestations.
3. **Mites (*Aceria tulipae*):**
    - **Symptoms:** Twisted leaves with yellow or light green streaking.
    - **Control:** Fumigate with methyl bromide after harvesting. Ensuring proper field sanitation and timely fumigation can reduce mite infestations.

Physiological Disorders

1. **Sprouting and Splitting:**
    - **Caused by:** Winter rains, excessive soil moisture, or nitrogen supply.
    - **Management:** Timely planting and controlled irrigation. Ensuring optimal soil moisture levels and proper timing of nitrogen applications can prevent these disorders.
2. Rubberification:
    - **Noticed in**: Rajkot and Nilgiri hills.
    - Controlled by: Micronutrients like zinc sulfate and ammonium molybdate. Regular soil testing and appropriate micronutrient applications can prevent rubberification.

Economic and Medicinal Value

Garlic is a significant export crop for India, valued for its culinary and medicinal uses. It enhances digestion, reduces cholesterol, and possesses antibacterial properties. Garlic extracts are used in traditional remedies for respiratory and digestive issues, reflecting its deep integration into cultural health practices. Additionally, garlic is known for its antioxidant properties and has been studied for its potential role in preventing certain types of cancers. Its economic value extends beyond direct sales, contributing to the health and wellness industry through various processed products and supplements. The versatility and benefits of garlic make it an indispensable crop in both agriculture and medicine.

Nutritional Composition of Garlic (*Allium sativum*)

Garlic, scientifically known as *Allium sativum*, is an essential bulb crop extensively utilized as a spice or condiment, similar to onions. It holds a significant place in both the culinary world and as a valuable commodity for foreign exchange in India, thanks to its substantial economic and health benefits. Rich in nutrients, garlic surpasses many other bulb crops in terms of its nutritive value. Its composition includes proteins, phosphorus, potassium, calcium, magnesium, and carbohydrates. Green garlic is especially high in ascorbic acid, enhancing its health benefits. The pungent aroma and flavor of garlic are attributed to the compound diallyl disulphide.

Garlic is not only a flavor enhancer but also a powerhouse of nutrition. Its proteins play a crucial role in various bodily functions, while its minerals support bone health, muscle function, and metabolic processes. The high potassium content helps regulate blood pressure, and the calcium is vital for strong bones and teeth. Magnesium in garlic aids in energy production and supports nerve and muscle function. Carbohydrates in garlic provide a quick source of energy, making it a valuable addition to diets, especially for those requiring an energy boost.

Detailed Nutritional Components

1. **Proteins:** Garlic contains essential amino acids that support muscle repair, growth, and overall cellular functions. These proteins are integral to the body's enzymatic activities and immune responses.
2. **Phosphorus:** This mineral is crucial for the formation of bones and teeth. It also plays a significant role in how the body uses carbohydrates and fats. Phosphorus is necessary for the body to make protein for the growth, maintenance, and repair of cells and tissues.
3. **Potassium:** Essential for maintaining fluid balance, nerve signals, and muscle contractions. Potassium helps counteract the effects of sodium, thus supporting cardiovascular health by helping to maintain normal blood pressure levels.
4. **Calcium:** Vital for bone health, calcium also supports heart, muscle, and nerve functions. Adequate calcium intake is crucial for preventing osteoporosis and maintaining overall skeletal strength.
5. **Magnesium:** This mineral plays a role in over 300 enzymatic reactions in the body, including the synthesis of proteins, muscle and nerve function, blood glucose control, and blood pressure regulation. Magnesium is also important for energy production and DNA synthesis.

6. **Carbohydrates:** These provide energy for the body's daily activities. The carbohydrates in garlic are primarily complex, offering sustained energy release and supporting digestive health.
7. **Ascorbic Acid (Vitamin C):** Found in high concentrations in green garlic, vitamin C is a powerful antioxidant that supports the immune system, skin health, and wound healing. It also enhances the absorption of iron from plant-based foods, contributing to better overall health.

Pungency Compound: Diallyl Disulphide

» **Diallyl Disulphide:** This sulfur-containing compound is responsible for the characteristic pungent aroma and taste of garlic. It has been studied for its potential health benefits, including anti-inflammatory, antioxidant, and antimicrobial properties. Diallyl disulphide plays a significant role in garlic's medicinal uses, contributing to its reputation as a natural remedy for various ailments.

Table: Nutritional Content of Garlic in Different Forms

Nutrient	Fresh Peeled Garlic Cloves	Dehydrated Garlic Powder
Moisture (%)	62.80	5.20
Protein (%)	6.30	17.50
Fat (%)	0.10	0.60
Mineral Matter (%)	1.00	3.20
Fibre (%)	0.80	1.90
Carbohydrates (%)	29.00	71.40
Energy (K. Cal)	145.00	--
Calcium (%)	0.03	0.10
Phosphorus (%)	0.30	0.42
Potassium (%)	--	0.70
Magnesium (mg/100g)	71.00	--
Iron (%)	0.001	0.004
Niacin (%)	--	0.70
Sodium (%)	--	0.01
Copper (mg/100g)	0.63	--
Manganese (mg/100g)	0.86	--

Zinc (mg/100g)	1.93	--
Chromium (mg/100g)	0.02	--
Vitamin A (IU)	0.40	175.00
Vitamin C (mg/100g)	13.00	12.00
Vitamin B (mg/100g)	16.00	0.68
Riboflavin B2 (mg/100g)	0.23	0.08
Thiamin (mg/100g)	0.06	--

Medicinal Value and Uses of Garlic

Garlic's health benefits are well-documented, primarily due to its high allicin content. Allicin is an odorless, water-soluble amino acid that becomes active when garlic is crushed, transforming into several volatile compounds, including the pungent diallyl disulphide. This compound is responsible for garlic's distinctive smell and many of its medicinal properties.

- » **Cholesterol Reduction:** Garlic has been shown to lower total and LDL cholesterol levels, which helps in maintaining cardiovascular health. The compounds in garlic inhibit cholesterol synthesis in the liver, contributing to reduced blood lipid levels.
- » **Immune System Boost:** Garlic enhances immune functions, making the body more resilient against infections and diseases. It stimulates the production of white blood cells, which play a critical role in the immune response.
- » **Antibacterial Properties:** The sulfur compounds in garlic have strong antibacterial and antiviral properties, making it effective against a variety of pathogens. Allicin, in particular, disrupts the production of essential enzymes in microbes, leading to their destruction.
- » **Anti-inflammatory Effects:** Garlic contains compounds that have anti-inflammatory properties, which can help in managing conditions like arthritis and other inflammatory diseases. These properties help reduce inflammation and associated pain.
- » **Traditional Medicine:** Garlic is used in traditional medicine to treat a wide range of ailments, from rheumatism to respiratory issues. It is believed to improve blood circulation, aid digestion, and detoxify the body.
- » **Antioxidant Properties:** Garlic's antioxidants help protect the body from oxidative damage, which can contribute to chronic diseases such as heart disease and cancer. These antioxidants neutralize free radicals, reducing the risk of cellular damage.

Garlic in Culture and History

Garlic is not only a culinary staple but also a significant part of cultural heritage across various civilizations. Believed to have originated in Central Asia and Southern Europe, particularly the Mediterranean region, garlic has been known as a culinary and medicinal plant since pre-dynastic Egypt, around 3000 BC.

- **Ancient Use:** Garlic was known to ancient Greeks and Romans, who valued it for its health benefits and culinary uses. It was used by soldiers and laborers to boost strength and endurance.
- **Spread to India and China:** Garlic spread to India and China, becoming integral to their culinary and medicinal practices. In India, garlic has been used in Ayurveda for its therapeutic properties, while in China, it has been a staple in traditional Chinese medicine.
- **Cultural Significance:** Despite its pungency, which was initially disliked by the Romans, garlic has been embraced globally. It symbolizes various cultural beliefs and traditions, such as being used to ward off evil spirits and in folk medicine for its perceived health benefits.
- **Modern Day:** Today, garlic is cherished worldwide for its flavor and health benefits. It remains a key ingredient in numerous cuisines and continues to be explored for its medicinal properties in scientific research.

6. Okra (*Abelmoschus esculentus* L.)

Botanical Classification:

Common Names: Lady's Finger, Bhindi

Botanical Name: *Abelmoschus esculentus* (L.) Moench

Family: Malvaceae

Chromosome Number: 2n = 130

Overview

Okra (*Abelmoschus esculentus*) is predominantly cultivated in tropical and subtropical regions. It ranks fifth in vegetable cultivation area in India, producing 33.24 lakh metric tonnes from 3.47 lakh hectares. This crop is highly valued for its young, tender fruits, which are commonly used in cooking for curries and soups. Okra is rich in vitamins A and B, proteins, and minerals, and is an excellent source of iodine, making it beneficial for treating goiter. The fruit also helps with genito-urinary disorders, spermatogenesis, and chronic dysentery. Okra fruits can be dried or frozen for off-season use. Additionally, dried okra fruits contain 13-22% edible oil and 20-24% protein, suitable for refined edible oil production. The dry fruit skin and fibers are utilized in the manufacturing of paper, cardboard, and fiber, while the root and stem help clarify cane juice for jaggery preparation.

Okra's nutritional profile makes it a versatile and valuable crop. Its high vitamin content supports immune function and overall health. The proteins and minerals present in okra contribute to muscle growth and metabolic processes. Iodine, crucial for thyroid function, makes okra a significant dietary component in regions prone to iodine deficiency. The various ways to preserve okra, such as drying and freezing, ensure its availability throughout the year, enhancing food security. The industrial uses of okra by-products demonstrate its economic importance beyond food consumption.

Origin and Distribution

Okra originated in tropical and subtropical Africa. The existence of many related species with significant variability suggests India as a secondary center of origin. India is the largest okra producer globally. Besides India, okra is also popular in Brazil, West Africa, and many other countries. In India, the major okra-producing states include Uttar Pradesh, Bihar, and West Bengal.

Okra's wide distribution highlights its adaptability to various climatic conditions. The genetic diversity found in its wild relatives provides valuable traits for crop improvement. As a staple vegetable, okra contributes significantly to the diets of many people in different regions, providing essential nutrients and culinary versatility. The major producing states in India benefit economically from okra cultivation, supporting livelihoods and local economies.

Taxonomy and Botany

Okra belongs to the Malvaceae family and was initially classified as *Hibiscus esculentus* L. due to its fused calyx, corolla, and staminal column, which fall off at anthesis. However, it was renamed *Abelmoschus esculentus* L. as its calyx is persistent. Okra is an annual herb with a lifecycle of 90-100 days. The flowers are bisexual and predominantly cross-pollinated, with anthesis occurring between 8:00-10:00 a.m. Anthers dehisce 15-20 minutes after anthesis, completing the process in 5-10 minutes. Pollen fertility is highest around one hour before and after flower opening, and fertilization occurs within 2 to 6 hours after pollination. The stigma is receptive at flower opening, making bud pollination ineffective in okra. The fruit is a capsule, and fiber development starts around the fifth to sixth day after pollination.

Most cultivated okra varieties are amphidiploids with 2n = 130 chromosomes. The chromosome number of *A. esculentus* ranges from 72 to 144, indicating its chromosome polymorphism. *A. esculentus* (2n = 130) evolved from crossing *A. tuberculatus* (2n = 58) with *A. ficulneus* (2n = 72), and colchicine treatment was used to make it an amphidiploid. Similarly, *A. caillei*, a complex polyploid, can be developed by colchicine treatment of the F1 cross between *A. manihot* and *A. esculentus*. The genus *Abelmoschus* includes both cultivated and wild forms, with species such as *A. esculentus*, *A. manihot*, *A. caillei*, and *A. moschatus*.

Understanding the taxonomy and botany of okra is crucial for crop improvement and breeding programs. The detailed knowledge of its reproductive biology helps in developing strategies for controlled pollination and hybridization. The genetic diversity within the genus

Abelmoschus provides opportunities to introduce desirable traits such as disease resistance, drought tolerance, and improved nutritional content into cultivated varieties.

Crop Improvement

Yellow Vein Mosaic Virus (YVMV) is a significant problem in okra cultivation. To address this, extensive efforts have been made to develop high-yielding and YVMV-resistant varieties. The development of the YVMV-resistant variety, Pusa Sawani, marked a significant advancement, reducing the importance of primitive, low-yielding local cultivars. After the breakdown of Pusa Sawani's resistance, research efforts intensified, leading to the development of numerous YVMV-resistant varieties.

The ongoing research and development in okra breeding aim to enhance not only disease resistance but also other agronomic traits. Breeders focus on improving yield, fruit quality, and environmental adaptability. The success in developing YVMV-resistant varieties has significantly boosted okra production, ensuring stable yields and reducing losses due to viral infections. Continued efforts in crop improvement are essential to address emerging challenges such as climate change, pest resistance, and changing consumer preferences.

Methods Used in Crop Improvement

1. **Plant Introduction**
    - **Example:** Perkin's Long Green
    - **Description:** Plant introduction involves bringing a plant variety from its native habitat to a new area where it has not been grown before. This method is often used to introduce new genetic material with desirable traits such as disease resistance, higher yield, or improved quality. Perkin's Long Green, an okra variety, is an example of successful plant introduction, contributing to the diversification and improvement of okra crops.
2. **Single Plant Selection and Pure Line Selection**
    - **Examp**le: Pusa Makhmali, Salkeerthi
    - Description: Single plant selection involves selecting the best individual plants from a population based on desirable traits. These selected plants are then used to develop pure lines, which are uniform and stable. Pusa Makhmali and Salkeerthi are examples of varieties developed through this method, ensuring high-quality and consistent traits in the progeny.
3. **Selection from Bi-parental Crosses**
    - **Example:** Pusa Sawani (IC 1542 x Pusa Makhmali)
    - **Description:** This method involves crossing two parent plants with desirable traits and selecting the best offspring from the resulting population. Pusa Sawani,

developed from a cross between IC 1542 and Pusa Makhmali, showcases the benefits of combining traits from two different parent lines to produce a superior variety with enhanced resistance and productivity.

4. **Selection from Complex Crosses**
   - **Example:** sel 2 from (Pusa Sawani x Best 1) x (Pusa Sawani x IC 7194)
   - **Description:** Complex crosses involve multiple parent plants and several generations of selection to combine and refine desirable traits. An example is sel 2, developed from a complex cross between (Pusa Sawani x Best 1) and (Pusa Sawani x IC 7194), illustrating the intricate process of combining multiple genetic backgrounds to create a variety with exceptional characteristics.

Significant Varieties

» **Susthira:** Developed at Kerala Agricultural University, this variety belongs to *Abelmoschus caillei*. It is known for its robust growth and adaptability to local conditions, making it a significant addition to okra cultivation in the region.

» **Hisar Unnat:** Developed from a cross between Sel 2-2 and Parbhani Kranthi, this variety is noted for its high yield and resistance to common okra diseases, contributing to its popularity among farmers.

» **Varsha Uphar:** Resulting from a cross between Lam Selection 1 and Parbhani Kranthi, Varsha Uphar is valued for its superior fruit quality and high yield, making it a preferred choice for commercial cultivation.

» **Parbhani Kranti:** This variety was developed by crossing *Abelmoschus esculentus* cv. Pusa Sawani with *Abelmoschus manihot* ssp. *Manihot*. Known for its resistance to Yellow Vein Mosaic Virus (YVMV), Parbhani Kranti is a significant improvement over traditional varieties.

» **Punjab 7:** Developed from a cross between *Abelmoschus esculentus* cv. Pusa Sawani and *Abelmoschus manihot* ssp. *Manihot* cv. Ghana, Punjab 7 is appreciated for its high yield and disease resistance, enhancing its appeal to farmers.

» **Arka Abhay:** Resulting from a cross between *Abelmoschus esculentus* and *Abelmoschus manihot* ssp. *Tetraphyllus*, Arka Abhay is known for its resilience and high productivity, making it a valuable addition to okra varieties.

» **MDU 1 and Punjab 8 (EMS 8):** These varieties were developed through mutation breeding using gamma rays and EMS, respectively. Mutation breeding has enabled the creation of varieties with unique traits, such as improved disease resistance and higher yield.

Varieties of Bhindi (Okra)

Indian Institute of Horticultural Research (IIHR), Bangalore

» **Arka Abhay**

- **Special Features:** Developed through hybridization, backcrossing, and selection from *A. esculentus* x *A. tetraphyllus* ssp. *tetraphyllus*. Resistant to Yellow Vein Mosaic Virus (YVMV) and tolerant to fruit borer. Fruits are 5-ridged, dark green, and medium-long. Productivity is 10.5 t/ha.
- **Additional Details:** This variety's resilience to YVMV and fruit borer ensures high yield stability and quality. Its medium-long fruits make it suitable for fresh market consumption, providing consistent productivity even in challenging conditions. Farmers prefer it for its robustness and ability to withstand common pests and diseases, thereby reducing the need for chemical interventions. The dark green fruits are also visually appealing, increasing their market value.

» **Arka Anamika**

- **Special Features:** Developed through hybridization, backcrossing, and selection from *A. esculentus* x *A. tetraphyllus* ssp. *tetraphyllus*. Resistant to YVMV. Fruits are 5-ridged, green, and medium-long. Productivity is 11.5 t/ha.
- **Additional Details:** Known for its disease resistance and high productivity, this variety is favored by farmers for its reliable yields and quality fruit. Its green, medium-long fruits are ideal for both fresh consumption and processing. The plant's vigorous growth and adaptability to various climates make it a popular choice in different agricultural zones. Additionally, Arka Anamika's resilience helps in minimizing crop losses, ensuring better economic returns for growers.

Indian Agricultural Research Institute (IARI), New Delhi

» **Pusa Makhmali**

- **Special Features:** Fruits are medium-long with a productivity of 8-10 t/ha. Susceptible to YVMV.
- **Additional Details:** Although it has high yield potential, its susceptibility to YVMV requires careful management to prevent disease outbreaks. The medium-long fruits are well-regarded for their texture and flavor, making them popular in culinary uses. Farmers need to employ integrated pest management strategies to protect this variety from viral infections. Despite its susceptibility, Pusa Makhmali remains a favored variety due to its high yield and market acceptance.

» **Pusa Sawani**

- **Special Features:** Fruits are 5-ridged, green, and medium-long. Productivity is 8-10 t/ha. Initially resistant but later became susceptible to YVMV.

- **Additional Details:** This variety was once a standard in okra cultivation due to its early resistance to YVMV, but the breakdown in resistance necessitates vigilant monitoring and disease management. Its green, medium-long fruits are still favored for their quality and taste. Pusa Sawani's history as a widely grown variety highlights the importance of ongoing research and breeding to maintain disease resistance.

» **Pusa A 4**

- **Special Features:** Plants have short internodes with 5-ridged, dark green, medium-long fruits and excellent shelf life. Productivity is 10-12 t/ha. Resistant to YVMV, jassids, and fruit borer.
- **Additional Details:** This variety's compact plant structure and disease resistance make it ideal for intensive cultivation and high-density planting. The dark green fruits have a longer shelf life, which is beneficial for transportation and storage. Pusa A 4's resistance to multiple pests and diseases ensures consistent yields and reduced crop losses. Farmers appreciate its reliable performance and economic benefits.

» **Selection 2-2**

- **Special Features:** Fruits are green, tender, long (16-20 cm), and 5-ridged.
- **Additional Details:** Known for its tender and long fruits, this variety is highly sought after for fresh markets and culinary uses. Its green, 5-ridged fruits are appealing to consumers for their taste and texture. Selection 2-2's productivity and quality make it a profitable choice for farmers. The long fruits are also ideal for various culinary applications, enhancing their market demand.

Indian Institute of Vegetable Research (IIVR), Varanasi

» **VRO 3**

- **Special Features:** Resistant to ELCV. Fruits are 5-ridged, green, and medium-long. Productivity is 11.2-16.5 t/ha.
- **Additional Details:** Its resistance to ELCV and high yield make it a reliable choice for farmers seeking disease-resistant varieties. The medium-long, green fruits are well-suited for both fresh consumption and processing. VRO 3's adaptability to different growing conditions contributes to its popularity. Farmers benefit from its consistent yields and reduced need for chemical treatments.

» **VRO 4**

- **Special Features:** Resistant to YVMV and ELCV. Fruits are 5-ridged, green, and medium-long. Productivity is 16.0 t/ha.
- **Additional Details:** This variety combines resistance to two major diseases with

high productivity, making it a valuable asset for okra cultivation. The 5-ridged, green fruits are attractive and marketable. VRO 4's robustness ensures minimal crop loss and stable economic returns. Its dual disease resistance makes it a preferred choice for sustainable farming practices.

Kerala Agricultural University

» **Kiran**

- **Special Features:** Shy branching plant with light green, long fruits (20-30 cm). Fruit weight is 25-30 g. Productivity is 11.21 t/ha.
- **Additional Details:** The long, light green fruits of Kiran are favored in markets, and its high productivity makes it economically viable for farmers. This variety's shy branching habit allows for easier management and harvesting. Kiran's fruits are also noted for their tender texture and flavor. Its adaptability to local growing conditions ensures reliable yields.

» **Salkerthi**

- **Special Features:** Attractive and long light green fruits (24-30 cm) with 5 ridges. Days to first harvest is 44. Productivity is 16.2 t/ha.
- **Additional Details:** Known for its early maturity and high yield, Salkerthi is ideal for quick turnover in the market. Its attractive, long fruits appeal to consumers and enhance marketability. The variety's early harvest time provides farmers with a competitive edge in the market. Salkerthi's consistent high yields make it a reliable income source.

» **Aruna**

- **Special Features:** Attractive red-colored long fruits (25-30 cm) with 7 ridges. Productivity is 15.8 t/ha.
- **Additional Details:** The unique red color of Aruna fruits makes them highly marketable and attractive to consumers. This variety's aesthetic appeal enhances its value in fresh produce markets. Aruna's long fruits are also prized for their taste and nutritional content. The high yield and distinct appearance make it a favorite among growers.

» **Susthira (A. caillei)**

- **Special Features:** Mosaic tolerant, late, and long-duration variety suitable for homesteads during the kharif season. Fruits are green and 22 cm long with 5 ridges. Productivity is 18 t/ha.
- **Additional Details:** Susthira's long duration and high productivity make it suitable for extended harvest periods and household consumption. Its tolerance

to mosaic virus ensures stable yields even in adverse conditions. The long, green fruits are ideal for fresh consumption and local markets. Susthira's robustness makes it a dependable choice for home gardeners.

Tamil Nadu Agricultural University (TNAU)

» **MDU 1**

- **Special Features:** Developed through gamma irradiation of seeds of Pusa Sawani. Plants are compact with closely arranged nodes. Fruits are 5-ridged, light green, and medium-long. Productivity is 10-11 t/ha. Susceptible to YVMV.
- **Additional Details:** Its compact plant habit makes MDU 1 suitable for high-density planting, though its susceptibility to YVMV requires effective disease management strategies. The light green fruits are well-received in the market for their quality. MDU 1's high productivity makes it economically attractive despite its disease vulnerability. Farmers need to implement integrated pest management to maximize yields.

» **CO.1**

- **Special Features:** Fruits are pinkish-red with a productivity of 12.0 t/ha. Susceptible to YVMV.
- **Additional Details:** The unique pinkish-red fruits of CO.1 add a distinct visual appeal, though careful management is needed to handle its YVMV susceptibility. This variety's attractive fruit color can command premium prices in the market. CO.1's productivity makes it a viable option for growers despite the need for vigilant disease control.

Haryana Agricultural University, Hisar

» **Varsha Uphar (HRB-9-2)**

- **Special Features:** Derived from Lam selection x Parbhani Kranti. Resistant to YVMV. Fruits are 5-ridged, dark green, and medium-long. Productivity is 9.8 t/ha.
- **Additional Details:** Varsha Uphar's resistance to YVMV and its productive capacity make it a reliable choice for growers. The dark green, medium-long fruits are appealing to consumers and suitable for various culinary uses. This variety's resistance ensures stable yields with minimal chemical interventions. Farmers appreciate its consistent performance and economic benefits.

» **Hisar Unnat (HRB 55)**

- **Special Features:** Plants are 3-4 branched, early, and high yielding (12-13 t/ha).
- **Additional Details:** Hisar Unnat's early maturity and high yield make it an excellent option for early market supply and increased profitability. The branching

habit promotes better air circulation and reduces disease incidence. This variety's early harvest time provides a competitive advantage in the market. Farmers benefit from its robust growth and high productivity.

Punjab Agricultural University, Ludhiana

» **Punjab 7 (P 7)**

- **Special Features:** Developed through the backcross method from *A. esculentus* cv. Pusa Sawani x *A. manihot* ssp. *manihot* cv. Ghana. Resistant to YVMV, jassids, and cotton bollworm. Fruits are 5-ridged, dark green, and medium-long with a deeply pigmented petiole base. Productivity is 9.5 t/ha.
- **Additional Details:** The resistance to multiple pests and diseases makes Punjab 7 highly desirable for sustainable cultivation. The dark green, medium-long fruits are visually appealing and marketable. Punjab 7's robust disease resistance ensures stable yields and reduces reliance on chemical treatments. Its pigmented petiole base adds to its distinct appearance.

» **Punjab Padmini**

- **Special Features:** Selection made at the F8 generation of a cross between the F1 of *A. esculentus* cv. Rashmi x *A. manihot* ssp. *manihot* with the F2 of *A. esculentus* cv. Pusa Sawani x *A. manihot* ssp. *manihot*. Fruits are dark green, 15-20 cm long, and five-ridged with pigmentation on the stem, petiole, and lower basal veins of leaves. Resistant to YVMV, jassids, and cotton bollworm. Yield is 10-12 t/ha.
- **Additional Details:** Punjab Padmini's high yield and multiple resistances ensure it remains a top choice for farmers facing diverse agricultural challenges. The dark green, five-ridged fruits are highly marketable. Its pigmentation adds a unique visual trait that enhances its market value. The variety's robust performance makes it a reliable income source.

» **Punjab 8 (EMS 8)**

- **Special Features:** An induced mutant from Pusa Sawani using EMS (1%). Fruits are 5-ridged, dark green, and medium-long. Field tolerance to YVMV and resistance to fruit borer. Fruits are green and 15-16 cm long.
- **Additional Details:** Punjab 8 combines mutation breeding with traditional selection to offer enhanced disease and pest resistance. Its medium-long fruits are suitable for both fresh markets and processing. The variety's field tolerance to YVMV ensures stable yields under challenging conditions. Punjab 8's high productivity makes it economically attractive for growers.

Marathwada Agricultural University (MAU), Parbhani

» **Parbhani Kranti**

- **Special Features:** Developed through the backcross method from *A. esculentus* cv. Pusa Sawani x *A. manihot* ssp. *manihot*. Fruits are 5-ridged, dark green, and medium-long. Productivity is 9.0 – 11.5 t/ha.
- **Additional Details:** Parbhani Kranti is noted for its robustness and high yield, making it a dependable variety for farmers. Its dark green, medium-long fruits are favored for their quality and taste. The variety's resistance to YVMV ensures minimal crop loss and consistent production. Parbhani Kranti's adaptability to various growing conditions enhances its appeal.

Chandra Shekhar Azad University of Agriculture and Technology (CSAUA&T), Kanpur

» **Azad Kranthi**

- **Special Features:** Fruits are 5-ridged, shining green with a long beak. Productivity is 12.5 t/ha.
- **Additional Details:** The distinctive shining green fruits of Azad Kranthi are highly marketable and favored for their appealing appearance. The long beak adds a unique visual trait that enhances its market value. This variety's high productivity makes it economically viable for growers. Azad Kranthi's robust performance ensures consistent yields and profitability.

Dr. Yashwant Singh Parmar University of Horticulture and Forestry (YSPUH&F), Solan, H.P.

» **Harbhajan Bhindi**

- **Special Features:** Perkins Long Green was released as Harbhajan Bhindi. Fruits are long, tapered green, and 8-ridged.
- **Additional Details:** The long, tapered fruits of Harbhajan Bhindi are ideal for fresh market sales and processing. This variety's robustness ensures high yield and quality fruit production. Its 8-ridged fruits are visually appealing and suitable for various culinary uses. Harbhajan Bhindi's adaptability to local growing conditions enhances its appeal to farmers.

Tamil Nadu Agricultural University (TNAU)

» **CO.2**

- **Special Features:** Cross between AE 180 x Pusa Sawani. Fruits are very long (22-25 cm), thick, 7-8 ridged, and light green in color. Productivity is 15-16 t/ha. Susceptible to YVMV.

- **Additional Details:** The exceptionally long fruits of CO.2 make it suitable for specialized markets, despite its susceptibility to YVMV. The thick, 7-8 ridged fruits are favored for their texture and taste. CO.2's high productivity makes it economically viable for growers. This variety's adaptability ensures reliable yields under various growing conditions.

» **CO.3**

- **Special Features:** Cross between Parbhani Kranti x MDU 1. Fruits are very long (22-25 cm), thick, 7-8 ridged, and light green in color. Productivity is 16-18 t/ha. Moderately resistant to YVMV.
- **Additional Details:** The variety's high yield and moderate resistance to YVMV make CO.3 an attractive choice for farmers seeking high returns. The long, thick fruits are ideal for fresh markets and processing. CO.3's adaptability to different growing seasons enhances its cultivation potential. This variety's robust performance ensures consistent production and profitability.

» **COBhH 1**

- **Special Features:** Cross between Varsha Uphar x Pusa A4 (T). Fruits are dark green, long, slender with five ridges. High yield – 22.1 t/ha. Resistance to yellow vein mosaic virus.
- **Additional Details:** COBhH 1's exceptional yield and disease resistance ensure its popularity among commercial growers. The dark green, long, slender fruits are highly marketable. This variety's robust performance ensures minimal crop loss and high economic returns. COBhH 1's adaptability to various growing conditions enhances its cultivation potential.

» **CO 1**

- **Special Features:** Pure line selection from Hyderabad "Red Wonder". Fruits are pinkish-red in color. Yield potential is 12 t/ha. Susceptible to yellow vein mosaic disease.
- **Additional Details:** The unique pinkish-red fruits of CO 1 are highly marketable, though its YVMV susceptibility requires careful management. This variety's high yield potential makes it economically viable. The distinctive fruit color adds value in fresh produce markets. CO 1's robust performance ensures reliable yields under favorable conditions.

» **CO 2**

- **Special Features:** F1 hybrid. Fruits are very long (22-25 cm), thick, 7-8 ridged, and light green in color. Yield potential is 15-16 t/ha. Susceptible to yellow vein mosaic disease. Suitable for preparation of dehydrated products.

- **Additional Details:** This hybrid variety's long fruits and suitability for dehydration add value to its cultivation. The thick, ridged fruits are favored for their texture and flavor. CO 2's high yield potential ensures economic viability for growers. This variety's adaptability to different growing conditions enhances its appeal.

» **CO 3 (Hy8)**

- **Special Features:** F1 hybrid between Prabhani Kranti and MDU 1. Yield potential is 16-18 t/ha. Moderately resistant to yellow vein mosaic. Suitable for growing during kharif as well as summer. Developed at the Department of Olericulture, Horticultural College and Research Institute, TNAU, Coimbatore.
- **Additional Details:** The hybrid's versatility for different growing seasons and moderate disease resistance make it a robust choice for farmers. CO 3 (Hy8)'s high yield potential ensures economic viability. The long, thick fruits are highly marketable and suitable for various culinary applications. This variety's adaptability ensures reliable production under varied conditions.

Acharya N.G. Ranga Agricultural University (ANGRAU)

» **Lam Hybrid Selection-1**

- **Special Features:** Plant height is 120-150 cm with occasional branching tendency. Fruits are slender, long, green, and five-ridged. Tolerant to yellow vein mosaic virus (YVMV).
- **Additional Details:** Lam Hybrid Selection-1's height and slender fruits make it suitable for varied growing conditions and market demands. The five-ridged fruits are appealing for fresh markets. This variety's tolerance to YVMV ensures stable yields and reduced crop loss. Farmers benefit from its robust performance and adaptability.

» **Janardhan**

- **Special Features:** Vigorous, branching plant. Fruits are small (6-8 cm long), suitable for canning and export. Tolerant to yellow vein mosaic.
- **Additional Details:** Janardhan's small fruit size and tolerance to YVMV make it ideal for processing and export markets. The vigorous branching habit promotes better air circulation and reduces disease incidence. This variety's adaptability ensures reliable yields under various growing conditions. Janardhan's robustness makes it a dependable choice for commercial cultivation.

Exploitation of Hybrid Vigor

Harnessing hybrid vigor, also known as heterosis, is key to improving Bhindi production. This concept involves the creation of hybrids that combine desirable traits such as increased seed production and larger flowers. Both public and private sectors have successfully developed hybrids resistant or tolerant to Yellow Vein Mosaic Virus (YVMV) using hand emasculation and pollination techniques.

Key hybrids include:

» **CO.2 (AE 180 x Pusa Sawani)**: This hybrid features exceptionally long fruits (22-25 cm) that are thick, have 7-8 ridges, and are light green in color. It yields 15-16 tons per hectare but is susceptible to YVMV. Despite this susceptibility, it is valued for its high yield and quality.

» **CO.3 (Parbhani Kranti x MDU.1)**: Similar to CO.2, this hybrid also produces very long fruits (22-25 cm) that are thick, with 7-8 ridges, and light green. It yields 16-18 tons per hectare and offers moderate resistance to YVMV, making it a more reliable choice for consistent production.

» **COBhH 1 (Varsha Uphar x Pusa A4 (T))**: Known for its dark green, long, slender fruits with five ridges, this hybrid has a high yield of 22.1 tons per hectare and is resistant to YVMV. Its robustness and productivity make it a preferred option for farmers.

These hybrids have significantly enhanced Bhindi production by combining beneficial traits from various parent lines, resulting in better yields and improved resistance to diseases.

Cultivation Practices

Effective cultivation practices are crucial for maximizing Bhindi yield and quality. These practices involve selecting the right soil, managing the climate, and ensuring proper sowing and spacing.

» **Soil**: Bhindi thrives in well-drained, fertile loamy soil with a pH of 6-7. Ensuring the soil is rich in nutrients and well-drained is essential for healthy plant growth. Regular soil testing and amendments help maintain optimal soil conditions.

» **Climate**: Bhindi grows best in tropical and subtropical regions, requiring a warm growing period with moderate rainfall. It cannot tolerate frost, and its performance suffers under drought, low night temperatures, and shade. High rainfall areas can boost growth and productivity, despite challenges in fertilization and seed set.

» **Sowing and Spacing**:

- *Sowing Time*:
  - North India: March-June (Spring-Summer), July-September (Kharif).
  - South and Central India: January-March (Summer), June-August (Kharif).
- *Spacing*:
  - Row spacing: 45-60 cm.

- Plant spacing: 30 cm.

Proper spacing and sowing timing are vital to ensure healthy plant growth and maximize yield. Adjusting sowing time and spacing can help mitigate the effects of adverse climatic conditions and pest infestations.

Climate

Bhindi is highly sensitive to climatic conditions. It cannot tolerate frost, and its performance declines under drought, low night temperatures, and shade. High rainfall can enhance plant growth and productivity but may hinder fertilization and seed set. Conversely, high temperatures and low humidity can stunt plant growth, resulting in shorter plants. Flowers are prone to dropping when daytime temperatures exceed 42°C. Managing these climatic factors is essential for ensuring successful Bhindi cultivation.

Land Preparation and Sowing

Effective land preparation is vital for successful Bhindi cultivation. The field should be ploughed thoroughly 2-3 times to achieve a fine tilth. Ridges and furrows or raised beds should be prepared, and seeds should be sown by dibbling on the sides of ridges or on raised beds. Soaking seeds for 6-12 hours before sowing can significantly enhance germination, especially during summer. This practice ensures a robust start for the crops, leading to better establishment and growth.

Manuring and Fertilizer Application

Proper manuring and fertilizer application are essential for providing Bhindi plants with the necessary nutrients for optimal growth. Applying 20-25 tons of farmyard manure as a basal dose during the final ploughing enriches the soil. A recommended dose of 150 kg each of N, P2O5, and K2O is used for Pusa Sawani in coordinated trials, though NPK recommendations can vary by region. For instance, in Kerala, a lower dose of 50:8:30 kg N is advised. One-third of the nitrogen, along with the full doses of phosphorus and potassium, should be applied as a basal dose. The remaining nitrogen should be applied in two split doses—four weeks after sowing and at the flowering and fruiting stages. Split applications of nitrogen after every third picking are beneficial for maintaining high yield and consistent fruit size.

Interculture

Maintaining a weed-free environment is crucial until the crop canopy fully covers the ground. This can be achieved through frequent hoeing, weeding, and earthing up. The use of weedicides like Lasso (2 kg a.i./ha), fluchloralin (1.5 kg/ha), or Metolachlor (1.0 kg a.i./ha) combined with one hand weeding at 45 days after sowing has proven effective and financially viable in All India Coordinated trials. Proper intercultural practices help in preventing weed competition, ensuring better growth and yield of the Bhindi plants.

Harvesting and Yield

Harvesting Bhindi fruits at the right time is essential to maintain their quality and market value. Fruits should be harvested when they reach maximum size but are still tender, typically when they are 6-8 cm long, which is ideal for export purposes. This usually occurs 5-6 days after flower opening. Harvesting should be done on alternate days using a knife or by bending the pedicel with a jerk. Cotton cloth hand gloves are recommended to protect fingers from the stinging effect. Harvesting in the morning, when fruit hairs are softer, and sprinkling water on pods at night to keep them cool and fresh for the market can help maintain quality.

» **Yield**:
  - Spring-summer crop: 6.0 – 8.0 t/ha.
  - Kharif crop: 10 – 12.5 t/ha.

Regular and timely harvesting ensures that the fruits remain tender and of high quality, which is crucial for both local markets and export. Employing best practices in harvesting can significantly impact the yield and quality of Bhindi, making it more marketable.

Post-Harvest Management

Effective post-harvest management is essential for maintaining the quality and marketability of Bhindi fruits. After harvesting, fruits should be graded and packed in jute bags, baskets, or perforated paper cartons. Sprinkling water on the fruits helps maintain their turgidity and prevents bruises, blemishes, and blackening. Pre-cooling of fruits before packing can further enhance their quality. Packed fruits should be transported in refrigerated vans for export, ensuring they remain fresh and appealing to consumers. Implementing proper post-harvest techniques helps in extending the shelf life and maintaining the quality of Bhindi, thereby meeting market standards.

Pests and Diseases

Diseases

» **Yellow Vein Mosaic Virus Disease (YVMV)**: This is the most severe disease affecting Bhindi, causing characteristic vein clearing and potentially leading to 100% yield loss if it occurs at an early stage. Affected fruits turn cream or white. The virus is transmitted by the whitefly (*Bemisia tabaci*). Control measures include removing susceptible weeds, controlling whitefly populations, uprooting and burying affected plants, adjusting sowing times, and cultivating resistant varieties like Arka Anamika, Arka Abhay, Susthira, and COBhH 1.

» **Cercospora Leaf Spot**: This disease thrives in high humidity and is common in seed crops. It causes sooty, black mold on the undersides of leaves, leading to leaf drop and blackish spots on mature pods. Control measures involve spraying Bavistin (0.1 g/l) or Bordeaux mixture at fortnightly intervals.

- **Powdery Mildew**: Caused by *Erysiphae chicoracearum*, this disease appears under prolonged humid conditions with white powdery pustules on the lower surface of leaves, resulting in yellowing and death of leaves. Control measures include spraying wettable sulfur (2 g/l) at fortnightly intervals.

Insect Pests

- **Jassids (Amrasca biguttula biguttula)**: These pests suck sap from the undersides of leaves, causing marginal yellowing, cupping, and drying of leaves, which leads to defoliation. A prophylactic spray of neem oil-garlic mixture at fortnightly intervals is recommended.
- **Fruit Borer**: Infestation causes toppling and death of young seedlings, withering and drying up of individual leaves and the central shoot, and severe fruit damage. Control measures include spraying carbaryl, thiodan, endosulfan, fenvalerate, cypermethrin, or deltamethrin, along with practicing summer ploughing and clean cultivation.
- **Nematodes**: Root knot nematodes infect roots, causing galls, premature leaf fall, wilting, and reduced growth and fruit production. Symptoms appear as well-defined patches in the field. Control measures include crop rotation with non-host plants like wheat, rice, and corn, successive deep ploughing during summer, soil solarization, and treating the field with nematicides.

Nutrient Management

Effective nutrient management is essential for optimizing the growth and yield of Bhindi (okra). The following practices are recommended:

- **Application of Organic Manure**: Incorporate 15-20 tonnes of well-decomposed farmyard manure per hectare into the soil. This organic matter enhances soil fertility, improves soil structure, and promotes beneficial microbial activity, which is vital for healthy plant growth.
- **NPK Fertilization**: The recommended application rates are 100 kg of nitrogen (N), 50 kg of phosphorus (P2O5), and 50 kg of potassium (K2O) per hectare. This balanced fertilization approach ensures that Bhindi plants receive adequate nutrients for vigorous growth, flowering, and fruit development. Regular monitoring of soil nutrient levels and adjustments based on soil tests can further optimize nutrient management.

Irrigation

Proper irrigation is crucial for the successful cultivation of Bhindi. Key irrigation practices include:

- **Regular Watering**: Ensure consistent irrigation, particularly during the critical flowering and fruiting stages, to support optimal plant growth and fruit set. Water stress during these stages can significantly reduce yield and fruit quality.

» **Avoid Waterlogging**: Excessive water and poor drainage can lead to waterlogging, which causes root rot and other root diseases. Implementing proper drainage systems and using raised beds can help prevent waterlogging, especially in areas prone to heavy rainfall.

Weed Management

Effective weed management is vital for reducing competition for nutrients, water, and light. Recommended practices include:

» **Regular Weeding and Hoeing**: Conduct frequent weeding and hoeing to keep the field free of weeds, particularly during the early stages of crop growth when Bhindi plants are most vulnerable to weed competition. Mechanical weeding or manual hoeing can effectively control weeds.

» **Mulching**: Apply mulch around the plants to suppress weed growth, conserve soil moisture, and regulate soil temperature. Organic mulches such as straw or leaves can also add organic matter to the soil as they decompose.

Diseases and Pest Management

Major Diseases

1. **Yellow Vein Mosaic Virus (YVMV)**:
    - **Symptoms**: Yellowing of veins and mosaic patterns on the leaves, leading to stunted growth and reduced yield.
    - **Management**: Utilize resistant varieties to mitigate the impact of YVMV. Additionally, controlling whitefly vectors through insecticides and cultural practices can help prevent the spread of the virus. Regular monitoring and timely intervention are crucial for effective management.
2. **Powdery Mildew (*Erysiphe cichoracearum*)**:
    - **Symptoms**: Presence of white powdery growth on leaves and stems, which can lead to leaf drop and reduced photosynthesis.
    - **Management**: Apply sulfur-based fungicides at the first sign of infection. Ensure good air circulation within the crop canopy to reduce humidity levels, which can help in preventing the disease.
3. **Cercospora Leaf Spot (*Cercospora abelmoschi*)**:
    - **Symptoms**: Circular to irregular brown spots on the leaves, which can coalesce and cause significant leaf drop.
    - Management: Use copper-based fungicides to control the spread of the disease. Remove and destroy infected plant debris to reduce the source of infection for subsequent crops.

Major Pests

1. **Jassids (*Amrasca biguttula biguttula*)**:
   - **Symptoms**: Yellowing and curling of leaves, leading to reduced photosynthetic activity and stunted growth.
   - **Management**: Apply insecticides such as Imidacloprid to control jassid populations. Regular monitoring and early intervention are key to preventing severe infestations.
2. **Whiteflies (*Bemisia tabaci*)**:
   - **Symptoms**: Yellowing and stunted growth due to sap-sucking, which also makes plants more susceptible to viral infections like YVMV.
   - **Management**: Use insecticides and sticky traps to reduce whitefly populations. Implementing reflective mulches can also deter whiteflies from landing on plants.
3. **Aphids (*Aphis gossypii*)**:
   - **Symptoms**: Curling and distortion of leaves, along with reduced growth and vigor.
   - **Management**: Apply insecticides such as Dimethoate to manage aphid infestations. Encouraging natural predators like ladybugs can also help keep aphid populations under control.

Medicinal and Economic Value

Bhindi is not only a nutritious vegetable but also possesses significant medicinal properties. It is used in traditional medicine to treat various conditions, including genito-urinary disorders, enhancing spermatogenesis, and managing chronic dysentery. Its mucilaginous properties are beneficial for soothing digestive issues.

Economically, Bhindi is a vital crop for small-scale farmers in tropical and subtropical regions. It provides a reliable source of income due to its high market demand and versatility in culinary uses. The consistent demand for fresh Bhindi, along with its relatively short growing cycle, makes it an attractive crop for farmers aiming to maximize their earnings from limited land resources. By following proper cultivation and post-harvest practices, farmers can significantly enhance the profitability and sustainability of Bhindi production.

Seed Production

Producing high-quality seeds for Bhindi (okra) requires careful planning and adherence to specific practices to ensure optimal yield and minimal disease incidence. The following guidelines are recommended for effective seed production:

» **Sowing Timing**: Adjust the sowing schedule to coincide with dry weather during the maturation and drying stages of the pods. This timing reduces the risk of diseases such as Yellow Vein Mosaic Virus (YVMV) and ensures the pods dry properly, which is crucial for high-quality seed production.

» **Isolation Distance**: Due to Bhindi's tendency to cross-pollinate, it is essential to maintain an isolation distance of at least 400 meters from other varieties. This distance prevents unwanted cross-pollination and ensures the genetic purity of the seeds being produced.

» **Field Inspection and Rouging**: Conduct regular field inspections and rouging at different stages of the crop's growth—pre-flowering, flowering, and fruiting. Removing off-type plants and those showing disease symptoms helps maintain the quality and uniformity of the seed crop.

» **Harvesting Practices**: Harvesting the initial two fruits is recommended to promote better plant growth and subsequent fruit set. This practice helps ensure that the plants have a robust structure and can support the development of high-quality seeds in later fruits.

» **Seed Yield**: Under optimal conditions, the average seed yield for Bhindi is approximately 1.0 to 1.5 tonnes per hectare. Ensuring proper agronomic practices and disease management can help achieve this yield consistently.

Variety	Special Features
MDU1	Developed at Agricultural College and Research Institute, Madurai, TNAU, by gamma irradiation of Pusa Sawani seeds. Light green fruits with long stylar end. Compact plants with closely arranged nodes. Susceptible to yellow vein mosaic disease. Yield 10-11 tonnes/ha.
Lam Hybrid Selection-1	Developed by ANGRAU. Plant height 120-150 cm with occasional branching. Slender, long, green fruits with five ridges. Tolerant to yellow vein mosaic virus.
Janardhan	Released by ANGRAU. Vigorous, branching plant. Small fruits (6-8 cm) suitable for canning and export. Tolerant to yellow vein mosaic.
Punjab Padmini	Developed at Punjab Agricultural University, Ludhiana. Dark green fruits weighing 20 g each. Tolerates yellow vein mosaic to some extent.
CO 1	Pure line selection from Hyderabad "Red Wonder". Pinkish red fruits. Yield potential of 12 tonnes/ha. Susceptible to yellow vein mosaic disease. Developed at TNAU, Coimbatore.
CO 2	F1 hybrid developed at TNAU, Coimbatore. Very long fruits (22-25 cm), thick with 7-8 ridges, and light green. Yield potential of 15-16 tonnes/ha. Susceptible to yellow vein mosaic disease. Suitable for dehydrated products.
CO 3 (Hy8)	F1 Hybrid between Prabhani Kranti and MDU 1. Yield potential of 16-18 tonnes/ha. Moderately resistant to yellow vein mosaic. Suitable for kharif and summer. Developed at TNAU, Coimbatore.

7. Cucurbitaceous Vegetables

Cucurbitaceous vegetables, commonly known as 'Cucurbits,' belong to the family Cucurbitaceae. These vegetables are diverse and are used for various culinary purposes, including salads (e.g., cucumber), cooking (e.g., gourds), pickling (e.g., West Indian gherkin), desserts (e.g., muskmelon, watermelon), and preserved or candied (e.g., ash gourd). In India and other tropical countries, cucurbits occupy the largest area among vegetable crops. While cucurbits are generally low to moderate in nutrient content, there are notable exceptions such as bitter gourd, which is rich in vitamin C (96 mg/100g), and parwal and chow chow, which have significant calcium content (531 mg and 140 mg/100g, respectively). Additionally, cucurbit seeds are highly valued for their oil and protein content, which is comparable to legumes and rich in methionine.

General Features

Cucurbits exhibit several distinct features that contribute to their adaptability and utility in various agricultural settings:

1. **Long Tap Root System**: The tap root can extend up to 175-180 cm, with lateral roots confined to the top 60 cm. This makes crops like bottle gourd, ash gourd, and parwal particularly suitable for riverbed cultivation, where deep rooting can access water and nutrients from deeper soil layers.
2. **Branched Stem**: Cucurbit stems are typically 3-8 branched, prostrate or climbing, and can spread up to 9-10 meters, particularly in species like Cucurbita and Lagenaria. Some varieties, such as Cucurbita pepo, have short internodes and exhibit a bushy growth habit, which can be advantageous for certain cultivation practices.
3. **Leaves**: The leaves of cucurbits are generally simple and mostly 3-5 lobed with a palmate structure. In some species like Citrullus, the leaves can be pinnately lobed, adding to the diversity within the family.
4. **Tendrils**: Tendrils are present in the axils of leaves and vary among species. They are simple in Cucumis, simple or bifid in other genera, and absent in bush types. Tendrils aid in the climbing habit of many cucurbit species, helping them secure themselves to various supports.
5. **Pollination**: Cucurbits are highly cross-pollinated, primarily by honey bees and bumble bees. The flowers are unisexual, large, and showy, often borne in the axils of leaves, either solitary or in racemose clusters. This characteristic ensures effective pollination and fruit set, contributing to high yields.
6. **Fruit**: The fruit of cucurbits is an inferior berry known as a "pepo," characterized by a hard rind when mature. Fruits like ash gourd and pumpkin can be stored for extended periods, whereas others like cucumber and snake gourd have shorter shelf lives. Except for chow chow, all cucurbit fruits are many-seeded, providing a substantial seed yield.

7. **Seeds**: Cucurbit seeds are borne in parietal placentation. The edible portion varies among species, such as the placenta in watermelon, the endocarp in ash gourd, ridge gourd, and smooth gourd, and the pericarp with some mesocarp in muskmelon. These seeds are nutritious and often used in various culinary and medicinal applications.
8. **Propagation**: Most cucurbits are propagated by seeds, which is an efficient and common method. However, a few exceptions like parwal and coccinia are vegetatively propagated, requiring different cultivation techniques.
9. **Habit**: Cucurbits are predominantly annuals, completing their life cycle within a year. Exceptions include chow chow and coccinia, which are perennial and continue to produce over multiple seasons.
10. **Cucurbitacins**: These bitter compounds are found in various parts of the plant, especially in fruits and roots. Pollen grains also carry cucurbitacins. The bitterness is common in oriental pickling melon, cucumber, and bottle gourd, which can influence their culinary uses and require specific processing methods to reduce bitterness.
11. **Sex Forms**: Cucurbits exhibit a range of sex forms, adding to their reproductive diversity:
    - **Hermaphrodite**: Plants with only bisexual flowers, seen in the Satputia variety of ridge gourd.
    - **Monoecious**: Plants with both male and female flowers, common in most cucurbits.
    - **Andromonoecious**: Plants with male and bisexual flowers, seen in muskmelon.
    - **Gynomonoecious**: Plants with female and bisexual flowers, seen in cucumber.
    - **Gynoecious**: Plants with only female flowers, found in some cucumber lines.
    - **Trimonoecious**: Plants with male, female, and bisexual flowers.
    - **Dioecious**: Male and female flowers on separate plants, seen in parwal, coccinia, and kakrol.

Flowering

Cucurbit plants typically begin flowering between 30-45 days after sowing, following a distinctive sequence that features an alternating pattern of male and female flowers until fruit set occurs. Initially, the first 4-6 flowering nodes on the plant predominantly produce male flowers. This is followed by the development of female flowers on subsequent nodes. This sequence is crucial for ensuring successful pollination and fruit set. In crops where immature fruits are harvested, such as cucumbers and zucchinis, this alternating pattern of flower production is less noticeable because the focus is on harvesting young, tender fruits rather than allowing them to mature fully.

In crops like melons and pumpkins, the situation can be more complex. Even if female flowers are produced, they may not necessarily result in fully developed fruits. This can be due to various factors, including inadequate pollination or environmental stresses that affect fruit set and development. The balance of male and female flowers, as well as the timing of their appearance, is influenced by several factors, including plant genetics, environmental conditions, and nutrient availability. Moreover, in cucurbit crops grown for seed production, the number of fruits per vine is generally lower than in those grown for vegetable production. This is because seed crops require a more significant investment of resources into producing viable seeds rather than a large number of fruits. This lower fruit set in seed crops ensures that the plants can allocate sufficient nutrients and energy to developing high-quality seeds. Understanding these flowering patterns and their implications is essential for growers to optimize both fruit and seed production, ensuring high yields and quality.

Pollination

Pollination in cucurbit plants primarily occurs early in the morning, typically between 6-8 AM. This timing is crucial because the flowers of cucurbit plants are generally short-lived, often wilting by midday. Crops like cucumbers, pumpkins, muskmelons, and watermelons depend heavily on this early morning pollination window for successful fertilization. However, the timing of pollination can vary with environmental conditions, particularly temperature. High temperatures can shift the optimal pollination time, necessitating careful observation and management by growers.

For crops such as snake gourd and pointed gourd, pollination follows a slightly different pattern. These plants often undergo anthesis, the period during which a flower is fully open and functional, at night. As a result, pollination typically occurs early the next morning. This unique timing requires different considerations for pollination management compared to other cucurbits.

Pollination in cucurbits is primarily entomophilous, meaning it is facilitated by insects. Bees, beetles, and moths are the main pollinators, with bees playing the most significant role due to their efficiency and prevalence. The degree of cross-pollination in cucurbits ranges from 60-80%, making it crucial to maintain a healthy population of pollinators in and around the cultivation area. Ensuring a diverse and abundant population of pollinators can significantly enhance fruit set and yield. Growers must be mindful of factors that can impact pollination, such as pesticide use, which can harm beneficial insects. Implementing integrated pest management (IPM) strategies can help mitigate these risks. Additionally, maintaining floral diversity within and around the fields can attract and sustain pollinator populations, ensuring effective pollination and high crop yields.

Sex Modification

Cucurbits exhibit a predominantly monoecious flowering habit, meaning they produce both male and female flowers on the same plant. The sex ratio in these plants is influenced by various environmental factors and the levels of endogenous hormones. High nitrogen levels, long daylight hours, and elevated temperatures tend to favor the production of male flowers. This male-biased flowering can affect fruit set, as fewer female flowers are available for pollination and subsequent fruit development.

Ethylene, a naturally occurring plant hormone, plays a significant role in promoting the production of female flowers. Its levels can be influenced by the application of plant growth regulators such as ethylene itself, malcic hydrazide, gibberellic acid (GA3), Ethrel, TIBA (2,3,5-triiodobenzoic acid), and boron. These substances can modify the hormonal balance within the plant, shifting the sex expression towards more female flowers. For instance, treatments with ethylene or Ethrel can increase the number of female flowers, enhancing the potential for fruit set.

Gibberellins, another group of plant hormones, generally promote the development of male flowers and are antagonistic to the effects of ethylene and abscisic acid, which promote female flower production. This hormonal interplay is complex and can be manipulated to achieve desired outcomes in sex expression. For example, in gynoecious lines of cucumber, which naturally produce only female flowers, male flowers can be induced through the application of GA3. This allows for controlled breeding and seed production in these specialized lines.

a. Bottle Gourd (*Lagenaria siceraria*)

Common Names: Bottle Gourd, Lauki (Hindi)

Botanical Name: *Lagenaria siceraria* (Mol.) Standl.

Chromosome Number: 2n = 22

Family: Cucurbitaceae

Bottle gourd, known for its distinctive shape that resembles a bottle, is a widely cultivated vegetable in India, available throughout the year. The fruit's historical use as a container has given it its name. The young, tender fruits of bottle gourd are commonly used in cooking, as well as in the preparation of sweets and pickles. When mature, the fruits develop hard shells that are used to craft domestic utensils, water jugs, and floats for fishing nets. Nutritionally, bottle gourd is celebrated for its easy digestibility and its cooling effect on the body. It possesses diuretic and cardiotonic properties, making it a valuable vegetable for health-conscious diets. The pulp of the fruit serves as an antidote for certain poisons and aids in controlling constipation, night blindness, and cough. Traditional medicine utilizes a decoction made from its leaves to treat jaundice, and its seeds are employed in treatments for dropsy.

Origin and Botany

The bottle gourd is believed to have originated in tropical Africa and has since been domesticated across Asia, Africa, and the New World. It grows as a climbing annual plant and typically has a growth duration of 3.5 to 4 months. Its flowers, which are solitary and chalky white, bloom at night—a characteristic that enhances its unique botanical profile. The fruits of the bottle gourd vary widely in shape and size, but all retain the fleshy texture that makes them so versatile in culinary uses.

Varieties and Cultivation

Bottle gourd is cultivated extensively and comes in various varieties developed by different agricultural institutions in India, each with unique characteristics:

- **Arka Bahar (IIHR, Bangalore)**: Known for its straight fruits without a crooked neck, medium size (1.0 kg at marketable stage), light green and shiny appearance. It has a high yield potential of 40-50 tons per hectare.
- **Pusa Naveen (IARI, New Delhi)**: Features cylindrical, straight fruits, averaging 550 grams, ideal for various culinary uses.
- **Pusa Summer Prolific Long and Round**: These varieties offer long (40-50 cm) and round (15-18 cm diameter) fruits, catering to different market and culinary needs.
- **Narendra Rashmi (NDAU&T, Faizabad, UP)**: Notable for its moderate tolerance to common pests and diseases, with bottle-shaped fruits weighing about 1.0 kg and a yield of 30 tons per hectare.
- **Pusa Komal (Punjab Agricultural University, Ludhiana)**: Known for its oblong, medium-sized fruits with light green skin and early harvest potential, showing tolerance to Cucumber Mosaic Virus (CMV).

Each variety has been bred to optimize yield, disease resistance, and specific agronomic traits, making them suitable for varied climatic conditions and cultivation practices.

Heterosis in Cultivation

The concept of heterosis, or hybrid vigor, is well-explored in bottle gourd cultivation, with several F1 hybrids showing significant improvements in yield and fruit quality. For instance, hybrids like Pusa Meghdoot, Pusa Manjari, and Pusa Hybrid 3 demonstrate enhanced yield potentials and desirable agronomic traits. The hybrid 'Varad' (MGH 4) developed by Mahyco in Jalna stands out with its cylindrical, bright green fruits, each 40-45 cm long and weighing 600-750 g, yielding an impressive 60-65 tons per hectare.

Optimal Growing Conditions

Bottle gourd thrives in warm-season climates and is more tolerant of cool conditions compared to crops like muskmelon and watermelon, though it cannot withstand frost. It grows best in well-drained fertile silt loam soils, which is why it's often found thriving in river bed cultivation setups, benefiting from the deep tap root system that supports its growth.

Seasonal Cultivation

Typically grown during the summer and rainy seasons, bottle gourd can be cultivated year-round in regions with sufficient water availability. This adaptability makes it a staple in both rural and urban agricultural settings, providing continuous yield and supporting the livelihoods of numerous farmers.

Varieties

Developing Institution	Variety	Special Features
IIHR, Bangalore	Arka Bahar	Fruits are straight without crook neck, medium-sized (1.0 kg at marketable stage), light green, and shining. Yield: 40-50 t/ha.
IARI, New Delhi	Pusa Naveen	Cylindrical, straight fruits, free from crooked neck. Average weight: 550 g.
	Pusa Summer Prolific Long	Long fruits (40-50 cm).
	Pusa Summer Prolific Round	Round fruits (15-18 cm diameter).
	Pusa Sandesh	Round green fruits, medium-sized (600 g). Early variety, ready for picking in 55-60 days. Yield: 29-32 t/ha.
	Pusa Meghdoot	F1 hybrid between PSPL and Sel.2. Long, light green fruits.
	Pusa Manjari	F1 hybrid between PSPR and Sel.11. Round, light green fruits.
	Pusa Hybrid 3	F1 hybrid with green, slightly club-shaped fruits without neck. Yield: 42.5 t/ha.
NDAU&T, Faizabad, UP	Narendra Rashmi	Moderately tolerant to pumpkin beetle, powdery mildew, and downy mildew. Fruits are bottle-shaped with a shallow neck. Average weight: 1.0 kg. Yield: 30 t/ha.
Punjab Agricultural University, Ludhiana	Pusa Komal	Medium-sized, oblong fruits with light green skin. Tolerant to CMV, early harvest (70 DAS). Potential yield: 40 t/ha.
	Punjab Round	Round, light green, and shining fruits. Yield: 15.5 t/ha.
CSAUA&T, Kanpur	Kalyanpur Long Green	Long fruits with a somewhat pointed blossom end. Yield: 30 t/ha in 120 days.

	Azad Nutan	Early variety with long neck-free fruits weighing 1.5 kg.
MPAU, Rahuri	Samrat	Cylindrical fruits without crook neck, greenish-white, 30-40 cm long, weighing 700-800 g. Yield: 43 t/ha.
CBPUA&T, Pantnagar	Pant Sankar Lauki 1	F1 hybrid with medium long fruits. Released by Central Variety Release Committee (CVRC) in 1999.
TNAU, Coimbatore	CO.1	Pale green, bottle-shaped fruits with narrow neck and round bottom. Yield: 25-30 t/ha.

Land Preparation and Sowing for Bottle Gourd

The preparation of land for bottle gourd cultivation involves several meticulous steps to ensure optimal growth and yield. Initially, the land is ploughed to achieve a fine tilth, similar to the preparation for ash gourd. Furrows are then created at intervals of 2.0 to 3.0 meters, which helps in efficient spacing and water management. Following this, well-decomposed farmyard manure is incorporated into the soil, enriching it with essential nutrients.

Seeds are sown in the prepared furrows, maintaining a spacing of 1.0 to 1.5 meters between plants to ensure adequate space for growth and air circulation. For plants trained on a bower—a structure designed to support climbing plants—spacing is adjusted to 3.0 x 1.0 meters. This arrangement optimizes sunlight exposure and air circulation, critical factors for vigorous growth. In areas with sloping land, sowing is done in pits that accommodate 2-3 plants each, which helps in water retention and prevents soil erosion.

To enhance germination rates, seeds are soaked for 12-24 hours in water or treated with succinic acid (600 ppm) for 12 hours. This pre-sowing treatment helps break seed dormancy and ensures uniform germination across the field. The typical seed rate for bottle gourd ranges from 3 to 6 kilograms per hectare, depending on the planting density and expected germination rates.

Training and Pruning

Bottle gourd plants exhibit vigorous vegetative growth, making training and pruning essential practices for managing their development. Training the plants on a bower not only helps in utilizing sunlight efficiently but also facilitates agricultural operations and improves fruit quality. Initially, axillary buds are removed to encourage vertical growth until the vines reach the bower. Upon reaching the bower, the apical bud is removed approximately 10-15 centimeters below the bower to promote the spread of 2-3 branches across it. After the plant has formed 4-5 fruits, a second pruning is conducted, which involves trimming the vines to leave only 2-3 axillary buds on each primary vine. This strategy helps focus the plant's energy on fruit development rather than excessive vegetative growth. Additionally, removing older, yellow, and pale leaves near the bottom of the plant helps prevent disease and promotes better air circulation around the fruiting zones.

Interculture

Bottle gourd plants respond very positively to substantial applications of manures and fertilizers. The fertilizer regime and interculture operations used for bottle gourd are similar to those applied for ash gourd and bitter gourd. These practices include regular weeding, soil aeration, and the application of balanced nutrients throughout the growing period to maintain soil fertility and plant health.

Harvesting

Harvesting of bottle gourd is typically done when the fruits are still tender, about one-third to half-grown, approximately 10-12 days after anthesis. This stage is determined by gently pressing the fruit skin and observing the persisting pubescence, which indicates tenderness and peak flavor. Cylindrical-shaped fruits are particularly favored in the market for their aesthetic appeal and ease of cooking.

Harvesting usually begins 55-60 days after sowing and is conducted at intervals of 3-4 days to ensure continuous production. Care is taken to avoid injuring the vines and fruits during harvesting. A sharp knife is used to cut the fruits, leaving a small part of the stalk attached to reduce water loss and prolong shelf life. The average yield for open-pollinated varieties is about 20-25 tons per hectare, while F1 hybrids can produce 40-50 tons per hectare, showcasing the benefits of advanced breeding techniques in enhancing productivity.

b. Ridge Gourd and Smooth Gourd

Common Names: Ridge Gourd (Hindi: Ghia tori), Smooth Gourd (Hindi: Kali tori)

Botanical Names: *Luffa acutangula* Roxb. (Ridge Gourd),
Luffa cylindrica Roem. (Smooth Gourd)

Chromosome Number: 2n = 26

The genus Luffa, known for the species ridge gourd and smooth gourd, offers significant cultural and commercial value. The name "Luffa" itself is derived from the common use of its mature dried fruits as "loofahs"—the fibrous network within the fruits is used globally as bathing sponges and in various other household products such as scrubber pads, doormats, pillows, mattresses, and cleaning utensils. This versatile use highlights the unique physical properties of the Luffa fruits, which allow them to transition from a food source to a utility product.

Commercial and Home Cultivation

Both ridge gourd and smooth gourd are extensively cultivated across various regions for their immature fruits, which are favored in many culinary traditions. These gourds are grown both commercially and in homestead gardens, reflecting their popularity and ease of cultivation. The cultivation process typically focuses on the young fruits, which are harvested before they mature and develop the fibrous texture that is unsuitable for consumption but ideal for loofah production.

The farming of Luffa gourds is often integrated into small-scale agricultural systems, providing a dual output—food and fibrous material. This dual-purpose cultivation makes Luffa crops particularly valuable for farmers looking to maximize the usability of their harvests. Commercial growers, on the other hand, might specialize in producing either the vegetable for culinary use or the fibrous material for industrial products, depending on market demands.

Nutritional and Medicinal Benefits

The young fruits of the Luffa gourds are not only valued for their culinary versatility but also for their health benefits. They are known to be easily digestible and appetizing, making them a suitable dietary option for the general populace as well as for individuals recovering from illness or those with sensitive digestive systems. The presence of a gelatinous compound called luffein in Luffa fruits is noteworthy; although not extensively studied, it is one of the unique biochemical components that contribute to the gourds' health benefits.

In traditional medicine, tender Luffa fruits are often recommended for individuals suffering from malaria and other seasonal fevers. This usage points to the anti-inflammatory and fever-reducing properties of the fruits, although the exact mechanisms and efficacy of Luffa in these contexts may require further scientific validation.

Utilization of Mature Fruits

As the fruits of ridge gourd and smooth gourd mature, they lose their culinary appeal but gain value as a source of natural fiber. Once they are fully dried, the fruits are processed to remove the outer skin and seeds, leaving behind the tough, fibrous skeleton that is famously transformed into loofahs. These natural sponges are then used worldwide in a variety of applications, from personal care to household cleaning. The sustainable nature of Luffa as a biodegradable and eco-friendly alternative to synthetic sponges adds to its appeal in today's environmentally conscious market.

Varieties - Ridge Gourd

Developing Institution	Variety	Special Features
IIHR, Bangalore	Arka Sumeet	Fruits lush green, cylindrical, 55 cm long, 2.5 cm girth, and 380 g weight with prominent ridges. Delicate aroma. Days to first harvest: 50-55.
	Arka Sujat	Fruits lush green, cylindrical, medium long (35-45 cm), average weight 350g. Yield: 63 t/ha in 100 days.
IARI, New Delhi	Pusa Nasdar	Fruits club-shaped, light green, 15-20 fruits/plant. Yield: 15-16 t/ha.
GBPU&T, Pantnagar	Pant Torai-1	Fruits 5 cm long, club-shaped. Yield: 10 t/ha. More suitable for rainy season.

Tamil Nadu Agricultural University	CO.1	Fruits long (60-75 cm), first harvest in 55 days, 10-12 fruits/plant weighing 3-4 fruits/kg.
	PKM-1	Fruits dark green, average weight 300g. Yield: 28-30 t/ha in 160 days.
	CO.2	Fruits very long (90-100 cm), weighing 350-400g, green with shallow grooves. Duration: 120 days. Yield: 25 t/ha.
HARP, Ranchi	Swarna Manjari	Tolerant to powdery mildew. Fruits elongated, medium-sized, highly ridged, green with soft flesh. Days to first harvest: 65-70. Yield: 18-20 t/ha in 140-150 days.
	Swarna Uphar	Fruits elongated, medium-sized (200g), ridges with soft, weak flesh. Days to first harvest: 65-70.
Punjab Agricultural University, Ludhiana	Punjab Sadabhar	Fruits long, 3-5 cm thick, slightly curved. Yield: 10 t/ha.
KKV, Dapoli	Konkan Harita	Fruits dark green, 30-45 cm long, tapering at ends, 10-12 fruits/plant.
Bihar	Satputia	Hermaphrodite cultivar producing pale green small fruits in clusters. Yield: 20-25 t/ha.

Varieties - Smooth Gourd

Developing Institution	Variety	Special Features
IARI, New Delhi	Pusa Chikni	Early variety. Fruits smooth, dark green, cylindrical, 15-20 fruits/plant.
	Pusa Supriya	Fruits pale green, 15-20 cm long, straight, slightly curved at stem end, pointed distal end with long peduncle. Average fruit weight: 110 g. Yield: 10-11 t/ha.
	Pusa Sneha	Suitable for long distance transport. Fruits dark green, 20-25 cm long with hard skin and soft flesh. Yield: 12 t/ha.
MPAU, Rahuri	Phule Prajakta	Fruits medium green with dark green sutures. Yield: 15 t/ha.
BAC, RAU, Sabour, Bihar	Rajendra Nenua 1	Fruits long, greenish white, smooth, and thick. Resistant to fruit fly and fruit rot. Yield: 25 t/ha.

Climate and Soil Requirements for Ridge and Smooth Gourd

Ridge and smooth gourds thrive in warm climates, making them ideal for cultivation during the summer and rainy seasons. The optimal temperature range for their growth is between 25-27°C, which promotes healthy development and maximizes yield. These crops prefer fertile, well-drained loamy soils rich in humus, which enhances water retention and provides essential nutrients that support vigorous growth and fruit development.

Cultivation Practices

For successful cultivation, ridge and smooth gourds are planted during specific times of the year to align with their growth preferences. Seeds are typically sown in January-February for the summer season and in June-July for the rainy season. Planting in raised beds, furrows, or pits helps manage water distribution and facilitates easier plant management. Seed rates vary slightly between the two gourd types, with ridge gourd requiring 3.5-5.0 kg/ha and smooth gourd needing 2.5-5.0 kg/ha.

Due to their hard seed coats, it is advisable to soak the seeds overnight in water to soften the coat and promote uniform germination. The recommended spacing for effective growth under a bower or trellis system is a row-to-row distance of 1.5-2.5 meters and a hill-to-hill distance of 60-120 centimeters. For those opting for a pit system, a row-to-row spacing of 1.5-2.0 meters and a pit-to-pit distance of 1.0-1.5 meters are ideal.

Manure, Fertilizer, and Irrigation

Manure and fertilizer applications, as well as irrigation and intercultural operations for ridge and smooth gourds, follow the same guidelines as those established for bitter gourd. These practices ensure that the plants receive adequate nutrients and water, contributing to healthy growth and productivity.

Harvesting

Ridge and smooth gourds are usually ready for harvest approximately 60 days after sowing. It is crucial to pick the fruits while they are still immature and tender, about 5-7 days after anthesis, to ensure the best quality for consumption. Fruits that mature beyond this point become fibrous and are not suitable for culinary use. To prevent over-maturity, fruits are picked at regular intervals of 3-4 days. Harvested fruits should be handled gently and stored in baskets to prevent damage; they can be kept for 3-4 days in a cool environment to maintain freshness.

Yield Expectations

Both ridge and smooth gourds typically produce yields ranging from 7.5 to 15.0 tons per hectare, depending on cultivation practices, environmental conditions, and plant health.

Pests and Diseases Management

Effective pest and disease management is crucial to maintaining the health of ridge and smooth gourd crops:

1. **Fruit Fly (Bacterocera cucurbitae)**: A significant pest, particularly for cucurbits. Control methods include using protective coverings on developing fruits, employing light traps and poison baits, and utilizing sex attractants.
2. **Red Pumpkin Beetle (Aulacophora foveicollis)**: Known for attacking most

cucurbits and causing significant damage. Control methods involve deep ploughing and applying specific insecticides like Furadan 3G.

3. **Epilachna Beetle (Epilachna septima)**: A serious pest for cucurbits like bitter and snake gourd, known for causing defoliation. Control strategies include mechanical removal of pests and chemical treatments with Carbaryl or Metacystox.
4. **Aphids (Aphis gossypii)**: These pests suck sap from leaves, causing them to crinkle, and can transmit the mosaic virus. Effective control includes the application of Malathion or natural remedies like tobacco decoction.
5. **Leaf Hopper (Amrasca biguttula biguttula)**: Causes significant damage to bitter gourd during summer. Control measures include spraying neem oil mixed with garlic or chemical insecticides like Acetaf and Imidachloprid.
6. **Red Spider Mites (Tetranychus sp.)**: These pests damage leaves by causing white patches and webbing. Control methods include spraying with neem oil mixed with garlic, Kelthane, or Dicofol.
7. **Leaf Miner (Lyriomyza trifolii)**: Known for causing distinctive white twisting lines on leaves. Control methods involve using yellow sticky cards to trap adults, burning infested leaves, and applying neem oil mixed with garlic.

Major Diseases Affecting Ridge and Smooth Gourd Crops

Ridge and smooth gourds are susceptible to a range of diseases that can significantly impact their growth and yield. Effective management of these diseases is crucial for maintaining healthy crops and ensuring optimal production. Here's a detailed look at the major diseases and their control measures:

1. **Fusarium Wilt (Fusarium oxysporum)**:
   - » **Symptoms**: This disease causes wilting of the plant along with a brown discoloration of the vascular bundles, which can severely stunt growth and lead to premature death of the plant.
   - » Control: To manage Fusarium wilt, it's recommended to use resistant varieties and practice crop rotation to prevent the buildup of pathogen levels in the soil. These strategies help reduce the incidence of the disease and maintain the health of the crop.
2. **Collar Rot (Rhizoctonia solani) / Pythium Rot (Pythium sp.)**:
   - » **Symptoms**: These diseases cause dark brown, water-soaked lesions at the base of the stem, which can weaken and eventually kill young plants.
   - » **Control**: Effective control measures include treating seeds with Thiram to prevent fungal infection and applying soil drenches of fungicides like Ridomil or Carbendazim to protect the developing plants.

3. **Powdery Mildew (Sphaerotheca fuliginea)**:

   » **Symptoms**: Characterized by white powdery spots on the leaves, this fungal infection can lead to significant defoliation and reduced photosynthetic capability.

   » **Control**: Regular spraying with fungicides such as Karathane, Calixin, or Carbendazim can help manage powdery mildew. Ensuring good air circulation within the crop can also reduce the prevalence of the disease.

4. **Downy Mildew (Pseudoperonospora cubensis)**:

   » **Symptoms**: This disease presents as water-soaked lesions on leaves, which can severely affect leaf health and function.

   » **Control**: Control methods include removing and destroying affected leaves to reduce the spread of the disease and applying protective fungicides like Dithane M-45 to prevent further infection.

5. **Anthracnose (Colletotrichum lagenarium)**:

   » **Symptoms**: Causes water-soaked spots on fruits and vines, leading to significant damage and potential crop loss.

   » **Control**: Implementing clean cultivation practices, rotating crops, and using fungicides such as Indofil M-45, Benomyl, or Carbendazim can effectively manage anthracnose.

6. **Alternaria Blight and Fruit Rot (Alternaria cucumerina)**:

   » **Symptoms**: This fungal disease causes yellow spots on leaves that turn black as they age, affecting the overall health and vigor of the plant.

   » **Control**: Using disease-free seeds, practicing clean cultivation and crop rotation, and spraying fungicides like Indofil M-45 are recommended to control this disease.

7. **Mosaic Virus**:

   » **Symptoms**: Infected plants show mottling, curling, and twisting of leaves, which can stunt growth and reduce yield.

   » **Control**: Collecting seeds from healthy plants and treating seeds with hot air or hot water are effective strategies. Additionally, controlling insect vectors that transmit the virus and using virus-tolerant varieties can help manage the spread of mosaic virus

c. Bitter Gourd (*Momordica charantia* L.)

Common Names: Bitter Cucumber, Balsam Pear, Karela (Hindi)

Botanical Name: *Momordica charantia* L.

Chromosome Number: 2n = 22

Overview of Bitter Gourd

Bitter gourd, known scientifically as *Momordica charantia*, is a popular vegetable in South Indian states, particularly in Kerala. This vegetable is cherished not only for its distinctive bitter flavor but also for its nutritional and medicinal benefits. Cultivated primarily for its immature tuberculate fruits, bitter gourd is rich in vitamins and minerals, boasting a high vitamin C content of 88 mg per 100 g. This makes it an excellent addition to a healthy diet.

Culinary and Medicinal Uses

The unique bitter taste of bitter gourd makes it a favored ingredient in various culinary preparations across South India and other Asian cuisines. It is commonly used in stir-fries, soups, and even as a pickled condiment. Beyond its culinary use, bitter gourd has a long-standing reputation in traditional medicine. It is believed to aid in the treatment of a range of ailments including diabetes, asthma, blood disorders, and rheumatism. The fruit pulp is considered an antidote for certain poisons, and a decoction made from the leaves of the plant is traditionally used to cure jaundice, reflecting the plant's diverse medicinal applications.

Origin and Distribution

Bitter gourd is native to tropical Asia, particularly the Indo-Burma region, and has been widely cultivated throughout India, Indonesia, Malaysia, China, and tropical Africa. Its adaptation to various tropical and subtropical climates has made it a staple in the diets of many cultures within these regions, highlighting its global agricultural significance.

Botanical Characteristics

Within the genus *Momordica*, which belongs to the family Cucurbitaceae, there are several species found in India:

- **M. charantia**: The standard cultivated variety known for its edible, bitter-tasting fruit.
- **M. charantia var. muricata**: A wild variety characterized by a higher bitterness and smaller fruits.
- **M. dioica**: Commonly known as spine gourd or kakrol, appreciated for its unique texture and flavor.
- **M. cochinchinensis**: Known as the sweet gourd of Assam, this variety contrasts with the typical bitterness of other members.
- **M. balsamina**: Another species whose immature fruits are utilized either as vegetables or in pickled form.

Momordica charantia is a monoecious annual climber that typically reaches maturity in 100-120 days. The plant features palmately lobed leaves ranging from 5 to 9 lobes, yellow flowers, and pendulous, fusiform fruits that are ribbed and covered with tubercles. The bitterness, a signature trait of the fruit, is primarily due to the presence of an alkaloid known as momordicin.

Varieties

Developing Institution	Variety	Special Features
IIHR, Bangalore	Arka Harit	Fruits short, spindle-shaped, green with smooth regular ribs, moderate bitterness. Yield: 9-12 t/ha.
IARI, New Delhi	Pusa Vishesh	Selection from local collection, suitable for summer. Fruits glossy green, medium long, thick.
	Pusa Do Mausami	Fruits dark green, club-like with 7-8 continuous ribs. Fruit weight: 100-120 g. Yield: 12-15 t/ha.
	Pusa Hybrid 1	Fruits medium thick, long, glossy green. Yield: 20 t/ha in 120 days.
Kerala Agricultural University	Priya (VK1)	Extra-long green spiny fruits with white tinge at stylar end. Average fruit length: 39 cm, weight: 235 g. Productivity: 24.5 t/ha.
	Preethi (MC 4)	Medium-sized white fruits with spines. Average fruit length: 30 cm, weight: 310 g. Productivity: 15 t/ha.
	Priyanka	Large white spindle-shaped fruits with smooth spines, thick flesh, fewer seeds. Average fruit length: 25 cm, weight: 300 g. Productivity: 28 t/ha.
Tamil Nadu Agricultural University	CO.1	Fruits dark green, medium length (20-25 cm), weight: 100-120 g. Yield: 14 t/ha.
	Coimbatore Long Green	Extra-long fruits (60 cm), dark green.
	MDU.1	Fruit weight: 300-450 g. Yield: 15-18 t/ha. Induced mutant with long (30-40 cm) greenish white fruits. Yield: 30-35 t/ha.
	Coimbatore Long White	Extra-long fruits (60-65 cm), white. Yield: 15 t/ha.
	COBgH 1	F1 hybrid between MC 84 x MDU1. Fruits light green, plump with more warts, each weighs 200-300 g. Yield: 44.4 t/ha in 115-120 days. Rich in momordicin (2.99 mg/100 g).
Konkan Krishi Vidya Peeth, Dapoli	Konkan Tara	Fruits green, prickly, medium long (15-16 cm), spindle-shaped. Yield: 24 t/ha.

Punjab Agricultural University, Ludhiana	Punjab 14	Plants bushy with light green fruits. Average weight: 35 g. Yield: 14 t/ha.
CSAUA&T, Kanpur	Kalyanpur Baramasi	Fruits long (30-35 cm), light green, thin, tapering. Tolerant to fruit fly and mosaic. Yield: 20 t/ha in 120 days.
MPKV, Rahuri	Hirkani	Fruits dark green, 15-20 cm long, spindle-shaped with warts and prickles. Yield: 14 t/ha in 160 days.
	Phule Green	Fruits dark green, 25-30 cm long, prickled, tolerant to downy mildew. Yield: 23 t/ha in 160-180 days.

Climate and Soil Requirements for Bitter Gourd

Bitter gourd thrives best in warm climates where the temperature ranges from 25-30°C, which are optimal for its growth and flowering. While it can adapt to slightly lower temperatures and high rainfall areas, it is important to note that temperatures above 35°C increase the plant's susceptibility to viral infections, affecting overall health and productivity. During cooler conditions below 10°C, germination can be significantly hindered due to the hard nature of the seeds. For successful cultivation, well-drained, fertile sandy loam or silt loam soils are ideal as they provide the necessary support for root development and nutrient uptake.

Seasonal Planting

Bitter gourd planting schedules vary significantly with geographical and climatic conditions:

» **Hills**: Typically sown during April-May when temperatures begin to rise.

» **Plains (Early Season)**: In cooler states like Rajasthan and Bihar, sowing occurs from January-March.

» **Plains (Late Season)**: In regions with late and prolonged winters, planting is done from February-March.

» **Mild Winter Areas**: Capable of supporting year-round cultivation due to mild temperatures.

» **Kerala**: Intensive sowing schedules include January-February for the summer season, May-June for the kharif season, and September for the rabi season.

Land Preparation and Sowing

Bitter gourd's land preparation and sowing practices share similarities with cucumber cultivation but require specific modifications such as training to a bower or trees for support. In Kerala, for instance, pits of 60 cm diameter and 30-45 cm depth are prepared at a spacing of 2.0-2.5 x 2.0-2.5 meters. These pits are enriched with 20-25 tonnes per hectare of well-rotten farmyard manure and filled with topsoil up to three-fourths of their height. Seeds, typically 5-6 kg per hectare, are soaked overnight and then stored in moist cloth for germination before being sown in the prepared pits.

Irrigation Practices

Regular and frequent irrigation, particularly during the fruiting stage, is crucial. The irrigation frequency ranges from every 2-5 days, depending on the stage of growth. In regions like Kerala, the crop is irrigated at 3-4 day intervals initially, and then on alternate days during the fruiting phase to ensure adequate moisture and nutrient availability.

Manures and Fertilizers

The nutritional management of bitter gourd involves applying 20-25 tonnes per hectare of farmyard manure along with 70 kg of nitrogen, 25 kg of phosphorus (P_2O_5), and 25 kg of potassium (K_2O) per hectare. One-third of the nitrogen, all of the phosphorus, and half of the potassium are applied as a basal dose either before sowing or within 10-15 days post-sowing. The remaining potassium is applied 45 days after sowing. Fertilizers are typically applied in several splits at fortnightly intervals to optimize nutrient uptake.

Interculture Operations

Given that bitter gourd has shallow root systems, deep intercultural operations should be avoided. Regular hand weeding, hoeing, and light earthing up are recommended alongside fertilizer applications. Pruning excess lateral branches is necessary to allow plants to reach the bower height effectively. Bowery systems, usually made of bamboo poles and G1 wire at a height of 2 meters, facilitate efficient trailing of vines and can be used for up to three cropping cycles.

Plant Growth Regulators

The application of plant growth regulators like MH (50-150 ppm), CCC (50-100 ppm), Ethrel (150 ppm), silver nitrate (3-4 ppm), and boron (3-4 mg/ha) at the 2-leaf and 4-leaf stages can significantly increase the number of female flowers and overall yield. Soaking seeds in solutions of Ethrel or boron also contributes to higher yields.

Harvesting and Seed Production

Harvesting begins 55-60 days after sowing when the fruits are fully grown yet still young and tender, ensuring the seeds within are not hard. Proper harvesting techniques are crucial to prevent damage to the plant and ensure continuous fruit bearing. For seed production, fruits are allowed to mature until they turn yellow; the seeds are then extracted, cleaned, and dried under shade. Seed yields are generally between 200-250 kg/ha.

d. Ivy Gourd (*Coccinia grandis* (L.) Voigt)

Synonyms: *Coccinia indica* and *C. cordifolia*

Overview of Ivy Gourd

Ivy gourd, commonly referred to as kundru or tondli, is a vigorous climbing perennial recognized for its distinctive fruits and adaptive growth. The fruits are small, typically 4-5 cm long and 1-2 cm in diameter, with an ovoid or elliptical shape. They are smooth and bright green with stripes, transitioning to scarlet red as they ripen. The plant is distinguished by its long tuberous roots and deeply lobed, five-lobed leaves with serrated edges, making it an interesting addition to tropical gardens.

Distribution

Ivy gourd is extensively cultivated across tropical and subtropical regions, with a prominent presence in India, tropical Africa, Malaysia, other Southeast Asian countries, and China. In India, it flourishes particularly in states like Tamil Nadu, Karnataka, Kerala, Maharashtra, Gujarat, Andhra Pradesh, and West Bengal. It belongs to the genus *Coccinia*, which includes about 30 species, mostly native to Africa; however, *Coccinia grandis* is the species primarily cultivated for agricultural and medicinal purposes.

Origin and Genetic Information

Originating from the lush tropics of India and Southeast Asia, ivy gourd has been successfully domesticated in these regions. It is genetically a diploid species with 2n=24 chromosomes. The plant exhibits a distinct sexual dimorphism where the male has one pair of XY chromosomes, making it heterogametic, and the female has a pair of XX chromosomes. This chromosomal arrangement is crucial for breeding and propagation purposes and does not impact the plant's dioecism.

Genetics and Breeding

Despite its widespread cultivation, detailed genetic studies and breeding programs for ivy gourd are relatively sparse. The existing cultivars are primarily developed through traditional methods like clonal selection, spontaneous mutations, or selection from seedling progenies by local farmers. The fruits of ivy gourd exhibit considerable variation in shape, size, and stripe patterns. Intriguingly, some cultivars can produce fruits parthenocarpically in the absence of male flowers, while others may produce bitter fruits that are generally not consumed.

Uses

The culinary and medicinal applications of ivy gourd are well-recognized. Immature fruits are commonly used in various dishes such as sambar and stir-fries, and the tender shoots are also edible. The plant's leaves, stems, and roots are utilized in traditional medicine, specifically in Ayurveda, to treat ailments such as skin diseases, bronchitis, diabetes, and are known for their properties to reduce blood sugar levels.

Production Technology

Climate

Ivy gourd prospers in warm and humid conditions, thriving best in temperatures between 20-30°C. It has a versatile fruiting pattern; in Southern India, it can bear fruits almost all year round, peaking during the summer and monsoon seasons. However, in the northern plains, it enters a dormant phase during the colder months, resuming growth with the onset of warmer weather.

Soil

The optimal soil type for ivy gourd cultivation is well-drained sandy loam or silt loam, which facilitates good root development and adequate drainage. Heavy clay soils are less suitable due to their poor drainage properties. The ideal soil pH for healthy growth is around 6.5, as the plant does not tolerate highly acidic or alkaline soils well.

Planting

Ivy gourd is propagated vegetatively using semi-hardwood cuttings, typically 15-20 cm long with 5-6 leaves each. Planting usually occurs in basins or pits of about 60 cm in diameter, spaced 1.5-2.0 meters apart to ensure adequate room for growth. In regions with distinct planting seasons, such as India, planting occurs twice a year in June-July and February-March, aligning with the rainy and dry seasons. Maintaining a female-to-male plant ratio of about 10:1 is crucial for effective pollination and fruit set.

Manures and Fertilizers

About 20 tonnes of farmyard manure per hectare are incorporated at planting to enrich the soil. Fertilizer application includes 60-80 kg of nitrogen, 40-60 kg of phosphorus, and 40 kg of potassium per hectare, distributed annually before the onset of fruiting. Pruning after fruiting is critical to manage plant size and encourage new growth, typically leaving vines about 60-75 cm long.

Irrigation and Intercultural Operations

Regular irrigation is necessary, especially during the fruiting phase, with intervals of 10-12 days in summer. During the rainy season, additional irrigation may not be necessary. It's important to manage soil moisture carefully, particularly during flowering, to avoid waterlogging, which can severely affect plant health. Light hoeing and weeding are also essential to maintain soil condition and prevent weed competition.

Yield

Flowering begins about 50-60 days after planting, with an average yield ranging from 10-15 tonnes per hectare, depending on cultivation practices and environmental conditions. The harvesting period is crucial, with fruits picked while still young and tender to ensure the best quality and flavor.

e. Ash Gourd (*Benincasa hispida* Cong.)

Common Names: Ash Gourd, Wax Gourd, Petha (Hindi)

Botanical Name: *Benincasa hispida* Cong.

Chromosome Number: 2n = 24

Overview of Ash Gourd

Ash gourd, known for its culinary versatility and medicinal properties, is cultivated for both its immature and mature fruits. In its immature form, it is popularly used as a cooked vegetable in various dishes. Once mature, the fruit is used in confectionery, most notably in the preparation of 'Petha,' a famous delicacy well-known throughout India. Additionally, ash gourd holds a significant place in Ayurvedic medicine, where it is used in preparations such as 'Kooshmanda rasayana.' It is particularly noted for its benefits for individuals suffering from nervousness, making it a valuable component in dietary and medicinal applications.

Origin and Distribution

Ash gourd, scientifically known as *Benincasa hispida*, is believed to have originated in Asia, specifically in regions like Java and Japan. It is now cultivated extensively across the old world tropics and is somewhat less common in the new world tropics. In India, the crop is extensively grown in Uttar Pradesh and Delhi, primarily for the production of 'Agra petha.' It is also widely cultivated in southern states where it is utilized as a vegetable, demonstrating its adaptability and cultural significance in various culinary traditions.

Botanical Description

The genus *Benincasa* is monotypic, with *B. hispida* being the sole species within this classification. Interestingly, there are no known wild forms of this species, suggesting a long history of cultivation that may have led to the loss of wild variants. Ash gourd is a diploid plant with a chromosome number of 2n = 24, indicating its stable genetic composition over time.

This plant is characterized by its vigorous yet slow-growing nature and is typically a trailing annual. Its long tap root system makes it particularly suitable for riverbed cultivation, where it can efficiently access deep water resources. The plant's stems and other parts are covered with bristle-like hairs, which may help protect against pests and physical damage.

Ash gourd is monoecious, meaning it produces both male and female flowers on the same plant. The male flowers are notable for their long pedicels, while the female flowers feature densely haired ovaries and short peduncles, distinguishing them from their male counterparts. The flowers have a large, yellow corolla, adding a splash of color to the plant. The floral sex ratio is significantly skewed towards male flowers, with a typical ratio of 34 staminate (male) to 1 pistillate (female) flower.

Flowering in ash gourd occurs early in the morning, with anthesis (the period during which a flower is open and functional) spanning from 4:30 to 7:30 AM. Anther dehiscence, the release of pollen, takes place between 3:00 and 5:00 AM, ensuring that pollen is available once the flowers open. The stigma of the female flower remains receptive from 8 hours before to 18 hours after anthesis, which allows for a considerable window for pollination to occur.

Varieties

Developing Institution	Variety	Special Features
Kerala Agricultural University, Thrissur	KAU Local	Medium-sized oval to oblong fruits with high flesh thickness, fruit length 45-55 cm, length ratio 2.05, fruit weight 6.1-8.1 kg, productivity 28.2 t/ha.
	Indu	Medium-sized round fruits, tolerant to mosaic disease, average fruit length 24.3 cm.
Tamil Nadu Agricultural University	CO.1	Fruits round, average fruit weight 5-6 kg. Duration 140 days.
	CO.2	Fruits small and long spherical, average fruit weight 3.0 kg.
APAU, Hyderabad	APAU Shakthi	Fruits long and cylindrical, yield 30-35 t/ha in 140-150 days.
UAS, Bangalore	Karikumbala	Local cultivar where the fruits are covered with an ashy coat.
IIVR, Varanasi	IVAG.502	Fruits oblong with an average weight of 12-13 kg. Yield 30-35 t/ha.

In addition to these open-pollinated varieties, a few F1 hybrids like MAH 1 and MHAG 2 have been developed in the private sector.

Climate Adaptation of Ash Gourd

Ash gourd, also known as winter melon, thrives in warm, humid environments and is particularly sensitive to cold temperatures. The optimal climate for cultivating ash gourd features consistent humidity and high rainfall, which support its vigorous growth and productivity. The ideal temperature range for its development is between 24°C and 30°C. These conditions ensure the plant can effectively carry out photosynthesis and other growth-related processes, contributing to the health and yield of the crop.

Seasonal Cultivation Practices

Ash gourd is adaptable to various seasonal conditions, particularly in regions with mild winters where it can be grown year-round. In tropical parts of India, such as Kerala, the sowing of ash gourd typically begins in May, coinciding with the onset of the monsoon season. In Tamil Nadu, planting occurs slightly later, around June to July. Conversely, in the northern parts of India where winters can be more severe, ash gourd cultivation is concentrated during the warmer summer and rainy seasons, avoiding the frost-prone colder months.

Cultivation Techniques

The cultivation of ash gourd shares several agronomic practices with that of pumpkins. These include similar spacing between plants, fertilizer application rates, and irrigation needs. One notable cultivation tip for enhancing germination and fruit set involves using seeds that are 3-4 months old, as older seeds tend to perform better than freshly harvested ones. Ash gourd's relatively slow initial growth rate makes it suitable for intercropping with quicker-growing, short-duration crops such as cucumber and oriental pickling melon, optimizing the use of space and resources in the garden or farm.

Key Cultivation Details:

Seed Rate: The seeding rate varies based on the cultivation method and location. In Kerala, under the pit system, about 0.75 to 1.0 kg of seeds per hectare is typical. In contrast, in the northern parts of India, a furrow system is employed, requiring around 5.0 kg of seeds per hectare.

Harvesting Techniques

Ash gourd can be harvested at different stages of maturity based on market demand and intended use. Immature fruits are typically harvested about a week after flowering, with subsequent weekly harvests to meet continuous market supply. For longer storage, transportation, or seed extraction purposes, it is preferable to harvest the fruits once a waxy coating has developed on their surfaces, indicating full maturity.

Harvest Yields:

» **Mature Stage:** Yields can range from 10 to 15 tonnes per hectare when the fruits are allowed to fully mature.

» **Immature Stage:** When harvested at the immature stage, the yields are significantly higher, approximately 25 to 30 tonnes per hectare.

Seed Production Methodology

For seed production, it is crucial to allow the fruits to mature fully until the vines and the stalks of the fruit have dried out completely. To enhance seed viability, it is recommended to store the fruits for about 2 to 3 months before seed extraction. The seeds are extracted by cutting the fruits open, scooping out the seeds along with the placenta, and allowing them to ferment overnight. After fermentation, the seeds are washed under running water and dried to a moisture content of about 8%. This process ensures a higher germination rate, as freshly harvested seeds typically exhibit lower viability.

Seed Yield:

» **Seed Production:** Effective seed production practices can yield approximately 175 to 200 kg of seeds per hectare.

Fertilizer Application Strategy

Proper fertilization is critical for the successful cultivation of ash gourd. At the planting stage, it is recommended to apply approximately 10 kg of farmyard manure (FYM) per pit. Additionally, a balanced application of NPK fertilizer (100 g of a 6:12:12 mixture per pit) along with an extra 10 g of nitrogen per pit after 30 days of sowing supports robust plant growth and productivity. This targeted fertilization strategy helps in establishing a strong foundation for the plants, leading to better fruit development and overall yield.

8. Beet Root (*Beta vulgaris L.*)

Common Names: Garden beet, Beetroot

Botanical Name: *Beta vulgaris L.*

Family: Chenopodiaceae

Climate and Soil Requirements

Beet root is a resilient crop that flourishes in cool climates, although it can also tolerate warm weather conditions. Optimal development of the roots, including the best color, texture, and sugar content, is achieved under cooler temperatures. When temperatures rise significantly, it can lead to the development of zoning, where alternating light and dark red concentric rings appear within the root. Exposure to low temperatures ranging from 4.5°C to 10.0°C for about 15 days can trigger bolting, which is the premature development of a flowering stem. Beet root thrives in deep, well-drained loam or sandy loam soils with a pH between 6 and 7. Additionally, beet root is one of the few vegetables that can grow in saline soils, making it a versatile crop for various soil conditions.

Varieties of Beet Root

There are several varieties of beet root, each offering unique characteristics suited to different growing conditions and market preferences:

Detroit Dark Red

» **Special Features:** This variety produces perfectly round roots with smooth, uniform deep red skin and dark blood-red flesh that may have light red zoning. It has a growing duration of 80-100 days, making it suitable for longer growing seasons.

Crimson Globe

» **Special Features:** Known for its round to flat round roots, this variety features medium red outer skin and crimson red flesh without zonations. It matures relatively quickly, within 55-60 days, making it ideal for areas with shorter growing seasons.

Early Wonder

» **Special Features:** This early-maturing variety has flat globular roots with dark red skin and flesh, exhibiting light red zoning. Its quick growth cycle makes it a good choice for early market production.

Ooty-1

» **Special Features:** Adapted for the Nilgiris region, this variety produces round roots with blood-red flesh. It yields approximately 28 tonnes per hectare in 120 days, making it highly productive under suitable conditions.

Crosby Egyptian

» **Special Features:** This variety is characterized by its flat globe roots with dark purplish-red flesh. It is known to develop white zoning under warm weather conditions and matures in 55-60 days.

Other Varieties

» **Special Features:** Varieties such as Madhur, Ruby Queen, and Ruby Red are available from private seed companies, each offering specific traits that cater to different market demands and growing conditions.

Cultivation Practices

Land Preparation and Sowing

Beet root is typically grown as a winter crop in plains and as a spring-summer crop in hilly areas. In the plains, sowing generally occurs from September to November. The land is prepared to a fine tilth, ensuring it is loose and friable to promote healthy root development. Well-decomposed farmyard manure is incorporated during the final ploughing to enrich the soil. Seeds are sown at a depth of 2.5 cm in rows spaced 45-60 cm apart, with 8-10 cm between plants. The recommended seed rate is 5-6 kg per hectare to ensure optimal plant density.

Manures and Fertilizers

To achieve the best growth, organic manure should be applied at a rate of 25 tonnes per hectare, especially in sandy soils. For average soil conditions, it is recommended to apply 60-70 kg of nitrogen, 100-120 kg of phosphorus, and 60-70 kg of potassium per hectare. Half of the nitrogen, along with all of the phosphorus and potassium, should be applied as a basal dose during land preparation. The remaining nitrogen should be applied 30-45 days after sowing to support continuous growth. Beet root has a high boron requirement; deficiency in boron can lead to internal breakdowns such as black rot or dry rot, which can significantly affect yield and quality.

Aftercare

Thinning is a crucial practice for beet root cultivation due to the nature of multiple seedlings germinating from each seed ball. It ensures that each plant has enough space to grow without competition. Maintaining soil moisture is essential for germination and subsequent growth. During the summer, 5-6 irrigations are typically required, while in the winter, 3 irrigations are sufficient in North Indian plains. Light hoeing during the early growth stages helps keep the field free of weeds. Additionally, earthing up—covering the swollen roots with soil—protects them from exposure to sunlight, which can cause them to toughen.

Harvesting Techniques

Beet roots are generally ready for harvest 8-10 weeks after sowing, once they reach a diameter of 3-5 cm. The roots are pulled by hand, and the tops are removed before grading and marketing. The average yield ranges from 25 to 30 tonnes per hectare. Properly stored beet roots, maintained at 0°C and 90% relative humidity, can retain their quality for extended periods, making them suitable for long-term storage and transport.

Seed Production

Seed production of beet root is most successful in hilly regions at altitudes above 1400 meters. The crop requires exposure to low temperatures (4.4 to 7.7°C) for 6-8 weeks to initiate flowering. The root-to-seed method is commonly used, where roots are dug out in November-December, selected, and replanted in well-prepared fields at a spacing of 60 x 45-60 cm. An isolation distance of 1000 meters is necessary for certified seed production, while 1600 meters is required for breeder and nucleus seed production to avoid cross-pollination and maintain seed purity. The average seed yield is approximately 2.0 tonnes per hectare.

Pest and Disease Management

Common Insect Pests

» **Leaf Miners:** These pests tunnel through the leaves, causing significant damage that can reduce the plant's photosynthetic capacity.

» **Web Worms and Semi Loopers:** These caterpillars feed on the foliage, leading to reduced plant vigor and potential yield loss.

Common Diseases

» **Cercospora Leaf Spot:** This fungal disease causes dark spots on the leaves, reducing the plant's ability to perform photosynthesis efficiently.

» **Downy Mildew:** This fungal disease affects the foliage, causing it to become discolored and malformed.

» **Viral Diseases:** Various viral infections can affect beet root, leading to symptoms such as stunted growth, leaf mottling, and reduced yields.

Effective pest and disease management involves regular monitoring of the crop, timely application of appropriate fungicides or insecticides, and adopting good agricultural practices such as crop rotation and maintaining field sanitation to minimize disease pressure.

9. Radish (*Raphanus sativus* L.)

Common Names: Radish, Mooli (Hindi)

Botanical Name: *Raphanus sativus* L.

Chromosome Number: 2n = 2x = 18

Family: Brassicaceae

Radish is cultivated primarily for its tender, tuberous roots, which are highly valued in various culinary applications. These roots are typically consumed raw in salads, adding a distinct pungent flavor that enhances the overall taste profile. Additionally, radishes can be cooked and enjoyed as a vegetable, providing a versatile ingredient in many dishes. In North India, radishes are a popular filling for parathas, often enjoyed with curd for breakfast, making a nutritious and satisfying meal.

The nutritional benefits of radish are numerous. It has a cooling effect on the body and helps prevent constipation due to its high fiber content. Radish is also known to increase appetite and is more nutritious when cooked with its leaves. The young leaves are also edible and are commonly used as a vegetable. Radish is recommended for patients suffering from piles, liver troubles, jaundice, and other ailments due to its medicinal properties. The juice of fresh radish leaves acts as a diuretic and laxative, further enhancing its health benefits. Rich in vitamin C and essential minerals, radish contributes significantly to a balanced diet.

Origin and Historical Significance

Radish (*Raphanus sativus*) originated in Europe and Asia, with historical evidence suggesting its evolution from *Raphanus raphanistrum*, a widely recognized weed crop in Europe. This ancient lineage highlights radish's longstanding significance in human agriculture and its adaptation to various climates and soil conditions across the globe.

Botanical Characteristics

Radish can exhibit both annual and biennial growth habits, depending on the specific ecotype or cultivar. The plant's rosette leaves are lyrate and pinnatifid, with leaf size varying from 10 cm to 15 cm in small-rooted cultivars and up to 45 cm in large-rooted cultivars. The edible portion of the radish develops from both the primary root and hypocotyls, resulting in the characteristic tuberous structure.

The inflorescence of radish is racemose, bearing white flowers that eventually develop into siliqua fruits. The seeds are initially yellowish but turn reddish-brown as they age. Radish is primarily cross-pollinated due to sporophytic self-incompatibility, with pollination typically facilitated by honey bees and flies. Stigma receptivity is maintained for up to four days after anthesis, allowing for effective pollination and seed set. Self-pollination can be achieved through bud pollination, ensuring genetic diversity and robust plant populations.

Detailed Botanical Features

Radish plants demonstrate a range of botanical features that contribute to their successful cultivation and propagation. The primary root, which forms the tuberous edible portion, varies in shape, size, and color depending on the cultivar. Some common root shapes include cylindrical, conical, and globular, with colors ranging from white to red, pink, and purple.

The leaves of the radish plant are an important indicator of plant health and vigor. Lyrate leaves are deeply lobed, resembling a lyre, and pinnatifid leaves have deep, feather-like divisions. These leaves play a crucial role in photosynthesis, supporting the growth and development of the tuberous roots.

The flowers of the radish plant are arranged in racemose inflorescences, which means they are borne on short stalks along a central axis. This arrangement allows for efficient pollination by insects, particularly honey bees and flies. The flowers are typically white but can sometimes exhibit shades of pink or purple.

The siliqua fruit of the radish plant is a type of dry fruit that splits open when mature to release the seeds. This characteristic facilitates seed dispersal and the establishment of new plants. The seeds themselves are small, initially yellowish, and turn reddish-brown as they mature. These seeds contain the genetic material necessary for the propagation of the next generation of radish plants.

Cultivation and Agricultural Practices

Radish is a cool-season crop that thrives in well-drained, fertile soils with a pH range of 5.5 to 6.8. It requires consistent moisture for optimal growth, particularly during the early stages of development. Radishes can be grown in various soil types, including sandy loam and loam, provided they are well-drained.

The planting depth for radish seeds is typically around 1 cm, with a spacing of 2.5 to 5 cm between plants and 30 to 45 cm between rows. This spacing ensures adequate room for the roots to develop and minimizes competition for nutrients and water. Radish plants require regular watering, especially during dry periods, to maintain soil moisture and prevent the roots from becoming tough and woody. Fertilization should be done based on soil tests, with an emphasis on nitrogen and phosphorus for root development. Organic matter, such as compost or well-rotted manure, can also be incorporated into the soil to improve fertility and structure. Weed control is essential for radish cultivation, as weeds can compete with the crop for nutrients, water, and light. Hand weeding, hoeing, and mulching are effective methods for keeping weeds in check.

Harvesting and Post-Harvest Management

Radishes are typically ready for harvest within 3 to 6 weeks after sowing, depending on the variety and growing conditions. Harvesting should be done when the roots reach a desirable size, typically 2.5 to 5 cm in diameter. Delaying harvest can result in the roots becoming woody and less palatable. After harvesting, radishes should be washed and cooled promptly to maintain their crispness and flavor. They can be stored in the refrigerator for up to two weeks, although their quality is best when consumed fresh. For longer storage, radishes can be stored in a cool, moist environment at temperatures close to 0°C.

Varieties

Asiatic Varieties

Developing Institution	Variety	Special Features
IARI, New Delhi	Pusa Deshi	Roots are pure white, pungent, 30-35 cm long, tapering with green stem end; suitable for sowing from mid-August to mid-October in North Indian Plains.
IARI, New Delhi	Pusa Chetki	Suitable for sowing in hotter months, March to August; roots are medium long, stumpy, pure white, medium pungent; duration 40-45 days.
IARI, New Delhi	Pusa Reshmi	Roots mildly pungent, 30-35 cm long, white with green shoulder; duration 55-60 days.
IARI, Katrain	Japanese White	Roots 20-30 cm long, cylindrical, mildly pungent, white with green shoulder; yield 15-30 t/ha in 60-65 days; suitable for sowing October-December.
PAU, Ludhiana	Punjab Safed	Roots white, tapering, 30-40 cm long, mild in taste, free from forking.
PAU, Ludhiana	Punjab Pasand	Roots are long, white, and semi-stumped. Suitable for sowing in October-November.
PAU, Ludhiana	Punjab Agethi	Roots medium long, red-skinned at the top and white-skinned at the lower half; most suitable for sowing from April to August.
CSAU&T, Kanpur	Kalyanpur No.-1	Roots 22-23 cm long, white with green shoulder, and tapering.
IIHR, Bangalore	Arka Nishant	Roots long, marble white, mild pungency; resistant to pithiness, premature bolting, root branching, and forking.
TNAU, Coimbatore	CO.1	Roots milky white, 23 cm long, cylindrical; yield 9-10 t/ha in 45 days.

European Varieties

Developing Institution	Variety	Special Features
IARI, Katrain	Pusa Himani	Roots white, 30-35 cm long, sweet flavored; suitable for growing year-round except winter months in hills; yield 32 t/ha in 60-65 days.
IARI, Katrain	White Icicle	Roots pure white, thin, icicle-shaped, straight, tapered; duration 30 days.
-	Scarlet Globe	Roots bright red, round, small, flesh white; takes 25-30 days from sowing to harvest.
-	Scarlet Long	Roots long, tapering with red skin and white flesh.
IIVR, Varanasi	Kashi Sweta	A selection made from Chetki population; roots 25-30 cm long, straight, tapering with pointed tip; yield 45 t/ha in 30-35 days.

Climate Requirements

Radish thrives best in cool climates, with an optimal temperature range for growth and development of high-quality roots being between 10°C and 15.5°C. These conditions are ideal for ensuring the radishes develop a mild flavor and tender texture. While radishes are capable of tolerating higher temperatures, such conditions often result in the roots developing a stronger pungency, which might not be desirable for all consumers. During hotter periods, it is crucial to harvest the roots while they are still small and tender to maintain their optimal flavor and quality. The plant's response to temperature makes it an adaptable crop, suitable for various climatic regions, but it thrives particularly well in cooler environments where its roots can develop to their full potential without the stress of excessive heat.

Soil Preferences

Radish requires soil conditions that are loose and friable, which allow the tuberous roots to expand without resistance. The soil should be rich in organic matter to provide the necessary nutrients for vigorous growth. Well-drained loam or sandy loam soils with a pH between 6 and 7 are ideal. These soil types ensure that the roots can grow uniformly, avoiding deformities and achieving the desired shape and size. Furthermore, radish is one of the few vegetable crops that can tolerate saline soils, making it a versatile choice for areas where soil salinity might pose a challenge for other crops. The incorporation of organic matter, such as compost or well-rotted manure, enhances soil structure and fertility, promoting healthy root development.

Land Preparation and Sowing Techniques

Radish can be cultivated year-round in the plains of North India, with specific sowing times depending on the variety and climatic conditions. For temperate radish varieties that can tolerate frost, the optimal growing period is from September to January. In contrast, tropical radish types are typically sown from mid-September onwards. If sowing is delayed beyond November, there is a risk of premature bolting, where the plants flower and set seeds earlier than desired.

To prepare the land for radish cultivation, it should be ploughed to a fine tilth, creating a loose and friable soil structure that supports root expansion. Ridges of approximately 25 cm in height are formed, with furrows spaced 30-45 cm apart. Seeds are sown continuously along the ridges, with the spacing between rows and plants varying based on the variety. Indian tropical cultivars, which tend to take longer to mature and grow larger, require a row spacing of 45 cm. After sowing, the seeds are thinned to maintain a spacing of 6-8 cm between plants. European types, which mature in 25-30 days, are sown at closer spacings of 5-10 cm between rows and 3 cm between plants. The seed rate for radish varies from 10 kg per hectare for larger varieties to 12 kg per hectare for temperate European types. For round cultivars, seeds are sown on the soil surface and covered with a thin layer of soil, while seeds of larger cultivars are sown 1.5-3.0 cm deep. Sowing seeds in a phased manner ensures a continuous supply of roots throughout the growing season.

Irrigation

Adequate water supply is crucial for the successful cultivation of radish from sowing to harvest. The soil must be kept consistently moist to promote rapid germination and subsequent root development. Irrigation should be applied immediately after sowing to ensure the soil remains loose and moist, which is essential for healthy root growth. Restricting irrigation can lead to tougher, more pungent roots, making them less desirable for consumption and marketing. Regular watering is essential, particularly during dry periods, to prevent the roots from becoming woody and maintaining their tender texture.

Inter-culture Practices

Inter-culture practices in radish cultivation involve activities that support the growth and health of the plants. As radish plants grow, the epicotyls develop into tuberous roots, which have a tendency to bulge out of the soil. To protect these roots and maintain their quality, earthing up is performed. This practice involves covering the exposed roots with soil, which also helps in weed control. Proper weed management is essential to reduce competition for nutrients, water, and light, ensuring the radish plants have optimal growing conditions. Additionally, maintaining soil moisture and providing adequate nutrients through the growing season are critical components of inter-culture practices.

Harvesting Techniques

Radish roots are generally ready for harvest 25-55 days after sowing, depending on the cultivar. Timely harvesting is crucial, as delayed harvesting can result in roots becoming bitter and pithy, reducing their market value and consumer appeal. Harvesting is typically done manually, and it is advisable to provide a light irrigation before pulling out the roots to facilitate easier removal from the soil. After harvesting, the roots are washed thoroughly, bundled with a few leaves for freshness, and marketed promptly to retain their quality. Proper post-harvest handling ensures the radishes remain crisp and flavorful until they reach the consumer.

Yield Potential

The yield of radish varies depending on the variety and growing conditions:

- **European Varieties:** These varieties typically yield between 5 to 7 tonnes per hectare.
- **Indian Cultivars:** These cultivars are more productive, yielding between 15 to 20 tonnes per hectare.

Radish roots can be stored at room temperature for 2-3 days without significant quality deterioration. For longer storage, roots can be kept at 0°C with 90-95% relative humidity for up to 2 months, maintaining their freshness and nutritional value.

Seed Production

Seed production for radish varies based on the type of variety. Seeds of European varieties are usually produced in hilly areas, while tropical types are produced in plains. Radish, belonging to the Cruciferae family, is cross-pollinated primarily by bees. To ensure genetic purity, an isolation distance similar to that required for cabbage or cauliflower is necessary. Wild relatives, such as wild mustard, wild turnip, and wild radish, should be removed from the field to prevent cross-pollination. For seed production, roots left in situ, without harvesting, produce the maximum quantity of seeds. An additional earthing up during the flowering and fruiting stages helps prevent lodging of the plants. For producing quality seeds, selected roots are pulled out, and true-to-type roots are replanted. The usual practice is to trim ½ to ¾ of the lower root portion before replanting, but this can reduce yield. Minimizing root injury during replanting is advisable for high seed yield. The average seed yield ranges from 600 to 800 kg per hectare.

Pests and Diseases Management

Diseases:

- **Alternaria Blight (*Alternaria raphani*):** This fungal disease presents symptoms similar to those seen in cole crops. Effective control includes crop rotation and the use of fungicides.
- **White Rust (*Albugo candida*):** This disease produces white, powdery patches on the underside of leaves and deforms flowering shoots. Using resistant cultivars such as Arka Nishant and applying fungicides like mancozeb (0.2-0.3%) can manage this disease.

Pests:

- **Aphids (*Myzus persicae* and *Brevicoryne brassicae*):** Aphids are the most serious pests of radish, feeding on the plant sap and potentially transmitting viral diseases. Regular monitoring and timely application of insecticides or biological control agents can help manage aphid infestations.

10. Carrot (*Daucus carota* L.)

Family: Apiaceae
Hindi: Gajar

Carrot is a historically significant cool-season root vegetable known for its edible, fleshy roots. These roots are widely used in various culinary dishes such as soups, stews, curries, pies, pickles, and salads. A famous sweet dish made from carrots is 'gajar halwa.' The roots are rich in alpha and beta carotenes, containing 1890 µg per 100 grams, and have a high sucrose content, which is ten times that of glucose or fructose. Additionally, carrot leaves are nutritious, providing a good source of leaf protein used as animal fodder and for making poultry feed. Carrots also offer numerous health benefits. They are known to increase urine output, help eliminate uric acid, and have a cooling effect, making them beneficial for individuals suffering from gall stones, constipation, and heat-related issues. Purple and black varieties of carrots are used to make 'kanji,' a beverage that acts as a good appetizer. In France, the essential oil from carrot seeds is used to flavor liquors and various food products, demonstrating the plant's wide range of uses.

Origin

Carrots are believed to have originated in Afghanistan, as suggested by Mackevic in 1929, due to the extensive diversity in morphological and root characteristics found there. Significant variability in root characteristics also exists in India, indicating it as another potential center of origin. Carrot root colors range widely from colorless to light lemon, light orange, deep orange, light purple, deep purple, and nearly black, showcasing the plant's genetic diversity.

Botany

Carrot is either an annual or biennial herb with an erect or branched stem that can grow to heights between 30-120 cm. The primary edible part of the carrot is its fleshy taproot, composed of an outer cortex (phloem) and an inner core (xylem). During the second phase of growth, the plant produces a compound umbel inflorescence. The individual flowers in this inflorescence are bisexual, with white or yellow petals. The "carrot seed" is actually a fruit known as an indehiscent mericarp, containing a single seed. Two mericarps combine to form a schizocarp, which is the actual carrot fruit developed from a two-loculed ovary.

Root Development

The development of carrot roots is a critical aspect of its growth cycle. In most cultivars, the root initially elongates before increasing in diameter. By about 12-16 days after sowing, the taproot reaches its optimal length. Several factors, including varietal characteristics, the presence of un-decomposed organic matter, and physical obstructions in the soil, influence root shape and size. Temperature also significantly affects root shape and color. For instance, the Chantney variety produces longer, more slender roots at 18°C, whereas roots become shorter and thicker at 24°C. High temperatures and inconsistent water supply can cause deep depressions on the roots, resulting in rough surfaces. Excessive irrigation following a drought period can lead to the formation of white corky outgrowths and side roots.

Cultivation Practices

Climate

Carrot cultivation is significantly influenced by climatic and soil conditions, which affect the shape and color of the roots. Carrots are typically grown as a spring-summer and autumn crop in hilly regions and as a winter crop in the plains of North India. The optimal temperature for seed germination ranges from 7.2°C to 23.9°C, while the best temperature range for root growth and development is between 18.3°C and 23.9°C. Tropical carrot varieties can produce roots even at higher temperatures, around 25°C. In contrast, European types require a lower temperature range of 4.8°C to 10.0°C for 4-6 weeks during any stage of root development or post-maturation, whether in storage or in the field. For seed stalk formation, plants need to experience subsequent temperatures of 12.2°C to 21.1°C.

Soil

Carrots thrive in deep, well-drained, and friable soils that support proper root development. For early crops, sandy loam is preferred, whereas silt loam is ideal for achieving high yields. Long-rooted carrot cultivars perform best in light soils, while short, stump-rooted varieties with round or heart-shaped roots are more suitable for heavy soils. In hard soils, the roots can become rough and coarse. The ideal soil pH for carrot cultivation ranges from 6.6 to 7.1.

Land Preparation and Sowing

Proper land preparation is crucial for successful carrot cultivation. The soil should be ploughed to a fine tilth, with care taken to remove clods and stones and ensure adequate drainage. The land is then formed into ridges and furrows spaced 30-45 cm apart in small plots suitable for irrigation. Carrot seeds can be sown on ridges or flat beds either by dibbling in lines or broadcasting. To facilitate even seed distribution, the seeds are rubbed to remove fine hairs and mixed with fine sand before sowing. For line sowing, a small furrow about 1.5 cm deep is made at the top or on either side of the ridges using a finger or a stick. The seeds, mixed with sand, are dropped into the furrows by hand and lightly covered with soil. The seed rate varies from 6 to 10 kg per hectare, depending on the variety and sowing method. Once the seedlings are established, they should be thinned to maintain a plant-to-plant distance of 5-8 cm to prevent overcrowding, which can lead to deformed roots.

It is advisable to provide light irrigation immediately after sowing or to soak the seeds in water for 12-24 hours before sowing to speed up germination.

Manuring and Fertilization

For optimal growth, carrots require both organic and inorganic fertilizers. In addition to applying 20-25 tonnes of farmyard manure per hectare, a recommended fertilizer dose includes 40-50 kg of nitrogen (N), 40-50 kg of phosphorus pentoxide (P_2O_5), and 80-100 kg of potassium oxide (K_2O). The entire dose of farmyard manure, half of the nitrogen, and all of the phosphorus and potassium should be applied as a basal dose during the final land preparation. The remaining nitrogen is applied during the first hoeing to support continued growth.

Irrigation

Adequate water supply is crucial for the successful cultivation of carrot from sowing to harvest. The soil must be kept consistently moist to promote rapid germination and subsequent root development. Irrigation should be applied immediately after sowing to ensure the soil remains loose and moist, which is essential for healthy root growth. Restricting irrigation can lead to tougher, more pungent roots, making them less desirable for consumption and marketing. Regular watering is essential, particularly during dry periods, to prevent the roots from becoming woody and maintaining their tender texture.

Inter-culture Practices

Inter-culture practices in carrot cultivation involve activities that support the growth and health of the plants. As carrot plants grow, the epicotyls develop into tuberous roots, which have a tendency to bulge out of the soil. To protect these roots and maintain their quality, earthing up is performed. This practice involves covering the exposed roots with soil, which also helps in weed control. Proper weed management is essential to reduce competition for nutrients, water, and light, ensuring the carrot plants have optimal growing conditions. Additionally, maintaining soil moisture and providing adequate nutrients through the growing season are critical components of inter-culture practices.

Harvesting and Post-Harvest Management

Carrot roots are generally ready for harvest 70-120 days after sowing, depending on the variety and growing conditions. It is important to harvest the roots at the right time, as over-mature roots can become woody and less palatable. Harvesting is usually done manually by pulling the roots from the soil. Light irrigation before harvesting can make the process easier by softening the soil. After harvesting, the roots should be washed to remove soil and other debris. They are then sorted, graded, and packed for the market. Proper post-harvest handling, including storage at low temperatures (0°C) and high humidity (90-95%), can extend the shelf life of carrots and maintain their quality.

Pests and Diseases Management

Carrots are susceptible to several pests and diseases, which can significantly impact yield and quality.

Common Pests

» **Carrot Rust Fly (*Psila rosae*):** The larvae of this fly burrow into the roots, causing extensive damage. Preventative measures include crop rotation and the use of floating row covers.

- **Aphids (*Aphis gossypii*):** These sap-sucking insects can weaken plants and transmit diseases. Biological control methods, such as introducing natural predators like ladybugs, can be effective.

Common Diseases

- **Alternaria Leaf Blight (*Alternaria dauci*):** This fungal disease causes dark, necrotic lesions on the leaves. Fungicidal sprays and resistant varieties are effective control measures.
- **Powdery Mildew (*Erysiphe heraclei*):** Characterized by a white, powdery growth on the leaves, this disease can be controlled by maintaining good air circulation and using fungicides.

Seed Production

Carrot seed production varies depending on the type of carrot being cultivated. Tropical annual types are typically produced in the plains, while temperate biennial types are grown in the hills. There are two primary methods recommended for carrot seed production: the root-to-seed method and the seed-to-seed method.

Root-to-Seed Method

The root-to-seed method involves growing carrots to maturity, harvesting the roots, and then replanting them to produce seeds. This method is particularly advantageous for producing high-quality seeds, as it allows for the selection of true-to-type roots, ensuring genetic purity and desirable traits in the next generation of plants. However, this method has a significant drawback: a high incidence of root rot. To mitigate this issue, it is recommended to transplant the roots without cutting them, as cutting can increase the vulnerability to rot.

Seed-to-Seed Method

The seed-to-seed method involves growing carrots directly from seeds to maturity in one continuous cycle. This method is generally followed to achieve higher seed yields. It is less labor-intensive than the root-to-seed method and avoids the complications associated with root transplanting, such as root rot.

Cross-Pollination and Isolation

Carrot is a cross-pollinated crop, primarily due to protandry, where the male parts of the flower mature before the female parts, promoting cross-pollination. Honey bees are the main pollinators for carrots, facilitating the transfer of pollen between plants. Given that carrots are cross-pollinated, it is crucial to maintain an isolation distance of at least 1000 meters from other carrot varieties to prevent cross-contamination and ensure genetic purity of the seeds.

Flowering and Seed Development

The flowering process in carrots begins with the opening of umbels, which starts from the periphery and progresses inward, completing within 6-7 days. This synchronized flowering pattern ensures effective pollination and seed set. After pollination, the seeds develop within the umbels. The average seed yield from carrot plants ranges from 500 to 600 kg per hectare, depending on the variety and growing conditions.

Physiological Disorders

Cavity Spot

One of the common physiological disorders affecting carrots is Cavity Spot, which is primarily caused by a deficiency of calcium induced by an excess of potassium in the soil. This disorder manifests as cavities in the cortex of the carrot roots, affecting both the quality and marketability of the crop. Proper soil management and balanced fertilization can help prevent this disorder, ensuring healthy and high-quality carrot roots.

Medicinal and Nutritional Benefits

Carrots are not only a staple in culinary traditions but also possess significant medicinal properties. They are known to enhance urine output and help eliminate uric acid from the body. The cooling effect of carrots makes them beneficial for individuals suffering from gallstones, constipation, and heat-related ailments. The high content of α and β carotenes in carrots contributes to their health benefits, including improved vision and a boosted immune system. Carrot leaves, rich in leaf protein, are valuable as fodder and in preparing poultry feed, underscoring the plant's overall utility.

Varieties

Carrot varieties differ in temperature requirements, length, size, shape, and color of roots, and crop duration. They are classified into temperate and tropical types.

Tropical Varieties

Variety	Description
Pusa Kesar	Developed at IARI, New Delhi by selection from a cross between Local Red and Nantes Half Long. Roots are scarlet in color with a sufficiently red-colored central core compared to the yellow or white core in Local; roots stay one month more than the Local red without bolting; contains high carotene (38 mg/100 g).
Pusa Meghali	Selection made at IARI, New Delhi by crossing Pusa Kesar and Nantes. It has long orange-colored tapering roots with self-colored roots.
Hisar Gairic	Roots long (18.5 cm), tapering, light brick red in color, less fibrous with a thin self-coloring core. Yield 25-30 t/ha.

European Varieties

Variety	Description
Nantes Half Long	Developed at IARI Regional Station, Katrain; small, slim, rough, sweet cylindrical roots with a stumpy tail; yield 15-20 t/ha.
Early Nantes	Cylindrical roots with a small thin tail, orange flesh with self-colored core; duration 90-100 days.
Chantenay	Suitable for canning and storage; reddish-orange roots 11.5-15.0 cm long.
Chaman	Developed at SKUA&T, Srinagar; roots long, cylindrical, and semi-blunt; tolerant to cracking; yields 25-27 t/ha.
Pusa Yamadagini	Developed at IARI Regional Station, Katrain; roots 15-16 cm long, orange with self-colored core, slightly tapering and semi-stumpy with medium top; 9-10 t/ha in 90-100 days.
Ooty-1	Developed at Horticultural Research Station (TNAU), Udhagamandalam: roots are 25 cm long with deep orange color; yield 45-50 t/ha in 100-110 days.
Zino	Introduced from Germany in the Nilgiris hills of Tamil Nadu by the State Department of Horticulture; roots deep orange with self-colored core; duration 110-120 days.

11. Turnip (*Brassica rapa*)

Common Names: Turnip

Botanical Name: *Brassica rapa*

Family: Brassicaceae

Climate Requirements

Turnips are best suited for cool and moist climates, which provide the most favorable conditions for their growth. Ideal root development, characterized by optimal flavor, texture, and size, occurs at temperatures ranging from 10°C to 15°C. This temperature range allows the roots to grow uniformly without developing the fibrous and tough texture associated with hotter weather. Short day lengths and cooler weather are especially beneficial for root growth, as they prevent premature bolting and ensure that the roots develop properly. Premature bolting, which occurs when plants flower and set seed early, can be a significant issue in turnip cultivation. Bolting is typically triggered by long days and high temperatures, conditions that stress the plants and cause them to shift from vegetative growth to reproductive growth prematurely. When turnips bolt early, the roots often remain underdeveloped and of poor quality, making them unsuitable for market.

In hot weather, the roots of turnips become more fibrous, tough, and pungent. This change in texture and flavor can render the roots less palatable, reducing their market value and consumer appeal. Therefore, it's essential to manage planting times to avoid high temperatures, especially during the critical stages of root development. Asiatic types of turnip exhibit a greater tolerance to higher temperatures compared to their temperate counterparts. These varieties can endure warmer climates without significant adverse effects on root quality. However, even Asiatic types perform best under cooler conditions. Temperate varieties, on the other hand, strictly require cooler weather to thrive, emphasizing the need for careful planning of sowing times and selection of appropriate varieties based on the local climate.

Soil Preferences

Turnips are highly adaptable to various soil types, but they show the best performance in loam soils enriched with humus. The presence of organic matter in the soil enhances its fertility and structure, promoting healthy root development. Well-drained sandy loam soils are particularly ideal for turnip cultivation as they provide the right balance of moisture retention and drainage. Extremely light sandy soils are generally unsuitable for turnip cultivation because they tend to dry out quickly and lack the necessary nutrients to support robust plant growth. Similarly, heavy soils, which can become compacted, impede root expansion and lead to misshapen or stunted roots. These soil types can cause the roots to develop defects, reducing their quality and marketability.

The ideal soil pH for growing turnips ranges from 6.0 to 7.0. Within this pH range, nutrients are more readily available to the plants, and the soil environment is conducive to healthy root growth. Adjusting soil pH to fall within this range can be achieved through soil amendments such as lime or sulfur, depending on whether the soil is too acidic or too alkaline. Proper soil preparation is crucial for successful turnip cultivation. This involves ploughing the soil to a fine tilth, which creates a loose and friable texture that supports root expansion. Removing clods, stones, and other debris is essential to prevent any physical barriers that could impede root growth. Ensuring adequate drainage is also vital to avoid waterlogging, which can lead to root rot and other soil-borne diseases.

Cultivation Practices

Land Preparation and Sowing

The cultivation of turnips requires meticulous land preparation to ensure optimal conditions for seed germination and root development. The timing of sowing is critical and varies depending on the type of turnip and the regional climate.

- **Asiatic Turnips:** These varieties are sown from July to September in the northern plains of India. This period aligns with the monsoon season, providing ample moisture and favorable temperatures for germination and early growth.
- **European Turnips:** These are typically sown from October to December, taking advantage of the cooler winter months in the same region.

- **Hilly Regions:** In hilly areas, sowing generally takes place from July to September, coinciding with the cooler and moist conditions ideal for turnip growth.

Proper soil preparation involves ploughing the land to a fine tilth, ensuring that the soil is loose and friable. This process creates an optimal environment for root expansion and development. Removing clods, stones, and ensuring proper drainage are crucial steps in land preparation to prevent any obstructions that could hinder root growth.

Seeds are sown directly either in lines or on ridges. For line sowing, small furrows about 1.5 cm deep are created at the top or on the sides of the ridges. Seeds are mixed with fine sand to facilitate even distribution and then sown into these furrows by hand, before being lightly covered with soil. The spacing of seeds within rows is 5-7 cm, with rows spaced 30 cm apart. Thinning is done 10-15 days after germination to maintain an optimal spacing of 10-15 cm between plants. This practice ensures that each plant has enough space to develop fully, reducing competition for nutrients and water. The seed rate varies depending on the type of turnip, typically around 4-5 kg per hectare.

Manuring and Fertilization

Turnips require a substantial amount of organic and inorganic fertilizers to achieve optimal growth. The application of 20-25 tonnes of farmyard manure per hectare during land preparation significantly improves soil fertility and structure. In addition to organic manure, a balanced fertilizer regimen is essential for providing the necessary nutrients:

- **Nitrogen (N):** 70-100 kg per hectare.
- **Phosphorus (P_2O_5):** 50 kg per hectare.
- **Potassium (K_2O):** 50 kg per hectare.

Half of the nitrogen and all of the phosphorus and potassium should be applied as a basal dose before sowing. This initial fertilization ensures that the young plants have access to essential nutrients during the early stages of growth. The remaining nitrogen is applied in two split doses: the first during root formation and the second during root development. This staged application of nitrogen supports continuous growth and helps in developing well-formed roots.

Irrigation

Consistent and adequate irrigation is critical for turnip cultivation. Regular irrigation intervals, typically ranging from 8 to 15 days, are necessary to maintain proper soil moisture levels. The frequency of irrigation depends on the weather conditions, with more frequent watering required during dry periods. Proper moisture levels are essential for the optimal growth and development of turnip roots. Insufficient irrigation can lead to the roots becoming fibrous and tough, while excessive watering can cause waterlogging and root rot.

To ensure uniform moisture distribution, light and frequent irrigations are recommended.

This practice helps in maintaining a consistent soil moisture level, which is crucial for the healthy development of the roots. During the early stages of growth, it is particularly important to keep the soil moist to support seed germination and seedling establishment.

Inter-culture Practices

Inter-culture practices play a vital role in maintaining the health and productivity of turnip crops. Regular hoeing and weeding are essential to keep the crop weed-free and conserve soil moisture. Weeds compete with turnip plants for nutrients, water, and light, which can significantly reduce crop yield and quality. Two to three hoeings are generally necessary during the growing season to control weeds and promote better soil aeration. Earthing-up, performed during the second and third hoeings, involves mounding soil around the base of the plants. This practice helps in producing better quality roots by preventing exposure to sunlight, which can cause discoloration and toughness. Earthing-up also supports the plants and helps maintain a clean root zone. The use of pre-emergence herbicides, such as Tok E-25 (Nitrofen) at 2 kg per hectare, can effectively control weeds and reduce the need for manual weeding. This herbicide application should be carefully managed to avoid any negative impact on the turnip plants.

Harvesting and Postharvest Management

Turnips are typically harvested when they are fully developed and tender, usually when the roots are 5-10 cm in diameter. Timely harvesting is crucial to prevent the roots from becoming tough and fibrous, which can occur if left in the ground for too long. Harvesting is generally done in the evening to minimize stress on the plants and to take advantage of cooler temperatures. The harvesting process involves loosening the soil with a spade and pulling out the roots by grasping the tops. In flat-ground cultivation, the tops are removed close to the ground, and the roots are dug out with a spade. Light irrigation before harvesting can facilitate easier uprooting of the roots.

After harvesting, the roots are washed to remove soil and other debris. They are then bundled and prepared for the market. Proper postharvest handling is essential to maintain the quality and freshness of the turnip roots. The roots can be stored for 2-3 days under cool and moist conditions, or for 8-16 weeks at 0°C with 90-95% relative humidity. These storage conditions help preserve the texture and flavor of the roots, extending their shelf life.

Seed Production

Seed production for turnips varies depending on the type:

» **Asiatic Types:** Seeds are produced in the plains.

» **European Types:** Seeds are produced in hilly regions.

For seed production, selected roots are pruned and transplanted. Maintaining an isolation distance of 1,000-1,600 meters from related crops is crucial to avoid cross-contamination. To achieve good seed yield, an application of 84 kg of nitrogen and 50 kg each of phosphorus and potassium per hectare is recommended. Seeds are harvested when 70% of the pods turn light yellow. The average seed yield is 5-6 quintals per hectare.

Physiological Disorder

Whip Tail

Whiptail is a physiological disorder caused by molybdenum deficiency, often induced by excess potassium. This deficiency results in narrow, cupped leaves with chlorotic mottling, adversely affecting root growth. Whip Tail is commonly found in acidic soils. It can be managed by liming the soil to adjust the pH to around 6.5 and applying 1.2 kg of sodium or ammonium molybdate per hectare.

Varieties

Variety	Special Features
Early Milan Red Top	Deep flat roots with purplish red top and white underneath; pure white flesh, well grained, crisp, mildly pungent.
Golden Ball	Globe-shaped roots, medium size, smooth; bright creamy yellow skin, pale amber flesh, fine texture and flavor.
Punjab Safed 4	Early maturing variety with pure white, round, medium-sized roots with mild taste.
Purple Top White Globe	Large, nearly round roots with purplish upper part and creamy lower part; white, firm, crisp, mildly sweet flesh.
Pusa Chandrima	Medium to large roots, flattened globe to globular shape, smooth, pure white skin; sweet and tender flesh.
Pusa Kanchan	Selection from Asiatic and European varieties; red skin, creamy yellow flesh, excellent flavor, short leaf top.
Pusa Swarnima	Flattish round roots with creamy-yellow skin and pale amber flesh; suitable for growing June-October in hills, October-December in plains.
Pusa Sweti	White roots, early maturing (45-50 days), suitable for October sowing in plains.
Snow Ball	Early temperate type; medium-sized small top, erect cut leaves; round, smooth roots with pure white skin.

Chapter - 6

Spices

Introduction

Spices are an integral part of human culture, history, and cuisine. These aromatic substances, derived from various parts of plants, such as seeds, fruits, roots, bark, and flowers, have been used for centuries to flavor food, preserve perishables, and even as currency in trade. Their significance transcends culinary boundaries, influencing medicine, cosmetics, and cultural rituals. This comprehensive note delves into the world of spices, exploring their origins, uses, and impacts on society.

Historical Significance of Spices

The history of spices is as rich and diverse as their flavors. Spices played a pivotal role in ancient trade routes, including the famed Silk Road, connecting the East with the West. The quest for exotic spices such as black pepper, cinnamon, cloves, and nutmeg spurred exploration and led to the discovery of new lands.

In ancient Egypt, spices were used not only for culinary purposes but also in religious rituals and mummification processes. The Greeks and Romans held spices in high regard, using them for medicinal purposes and as status symbols. During the Middle Ages, spices became even more precious in Europe, often worth their weight in gold. The lucrative spice trade led to the establishment of trade empires by the Portuguese, Dutch, and British, fundamentally altering global economic and political landscapes.

The Botanical and Geographical Diversity of Spices

Spices are derived from various parts of plants, each with unique characteristics and flavors:

- **Seeds:** Examples include cumin, coriander, and mustard seeds, which are known for their intense flavors and essential oils.
- **Fruits:** Black pepper, allspice, and cardamom are derived from fruits, offering distinct tastes and aromatic qualities.

DOI: 10.1201/9781003790938-6

- **Roots and Rhizomes:** Ginger, turmeric, and galangal are notable examples, valued for their pungent flavors and health benefits.
- **Bark:** Cinnamon and cassia are derived from tree bark, providing warm and sweet flavors.
- **Flowers and Flower Buds:** Cloves, saffron, and capers are obtained from flowers or flower buds, contributing unique tastes and colors to dishes.

The geographical distribution of spices is as diverse as their botanical sources. Tropical regions, particularly in South and Southeast Asia, are the primary producers of many spices. India, for instance, is a major producer and exporter of spices such as black pepper, cardamom, turmeric, and cumin. Indonesia is renowned for nutmeg and cloves, while Sri Lanka is famous for its high-quality cinnamon. The climatic conditions in these regions, characterized by warm temperatures and ample rainfall, provide the ideal environment for spice cultivation.

Culinary Uses of Spices

Spices are the heart of global cuisine, transforming ordinary ingredients into extraordinary dishes. Their use varies from region to region, reflecting cultural preferences and local traditions.

- **Indian Cuisine:** Spices are the backbone of Indian cooking, used in complex blends known as masalas. Turmeric, cumin, coriander, and chili are fundamental to many dishes, while garam masala, a blend of several spices, adds depth and warmth.
- **Middle Eastern Cuisine:** Spices such as sumac, za'atar, and saffron are essential in Middle Eastern dishes. They enhance the flavors of meats, grains, and vegetables, creating aromatic and flavorful meals.
- **Asian Cuisine:** Chinese, Thai, and Japanese cuisines use spices like star anise, ginger, and wasabi to create distinctive tastes. Spices are often combined with soy sauce, fish sauce, and vinegar to balance flavors.
- **European Cuisine:** Historically influenced by the spice trade, European cuisine incorporates spices such as black pepper, nutmeg, and cloves. These spices are used in both savory dishes and desserts, adding a touch of warmth and complexity.
- **Latin American Cuisine:** Chili peppers, cumin, and cinnamon are staples in Latin American cooking. These spices contribute to the bold and vibrant flavors characteristic of the region's cuisine.

Medicinal Uses of Spices

Beyond their culinary applications, spices have been used for their medicinal properties for thousands of years. Traditional medicine systems such as Ayurveda, Traditional Chinese Medicine (TCM), and Unani heavily rely on spices for their therapeutic benefits.

- **Turmeric:** Known for its anti-inflammatory and antioxidant properties, turmeric is used to treat a range of conditions, including arthritis, digestive disorders, and skin diseases. Its active compound, curcumin, is the subject of extensive research for its potential health benefits.

- **Ginger:** Valued for its digestive and anti-nausea properties, ginger is commonly used to alleviate gastrointestinal discomfort, motion sickness, and morning sickness during pregnancy. It also has anti-inflammatory effects and can help reduce muscle pain and soreness.
- **Cinnamon:** This spice is known to help regulate blood sugar levels and improve insulin sensitivity. It is also used for its antimicrobial properties and to improve digestion.
- **Cloves:** Cloves possess strong antimicrobial and analgesic properties, making them useful for dental health. Clove oil is often used to relieve toothaches and oral infections.
- **Pepper:** Both black and long pepper are used to enhance digestion and stimulate appetite. Piperine, the active compound in black pepper, has anti-inflammatory and antioxidant properties.

Economic Impact of the Spice Trade

The spice trade has had a profound impact on the global economy throughout history. The demand for spices drove exploration and the establishment of trade routes, significantly influencing the development of early economic systems. In ancient times, spices were so valuable that they were used as currency and given as gifts to royalty and deities.

The Age of Exploration, spearheaded by European powers, was largely motivated by the quest for direct access to spice-producing regions. This period saw the establishment of colonial empires and the opening of new markets for spices. The Portuguese, Dutch, and British East India Companies played pivotal roles in controlling and monopolizing the spice trade.

Today, the global spice market continues to thrive, with major producers such as India, Vietnam, Indonesia, and China exporting vast quantities of spices worldwide. The spice industry supports the livelihoods of millions of farmers, traders, and processors, contributing significantly to the economies of producing countries.

Spices in Cultural and Religious Contexts

Spices hold significant cultural and religious importance across various societies. They are often used in rituals, ceremonies, and traditional practices, symbolizing prosperity, purity, and health.

- **India:** Spices are an integral part of Indian cultural and religious practices. Turmeric is used in Hindu ceremonies and is considered auspicious. Saffron is used in religious offerings and is believed to bring good luck and prosperity.
- **Middle East:** In Middle Eastern cultures, spices like saffron and cardamom are

used in hospitality rituals. Offering spiced coffee or tea to guests is a sign of respect and generosity.

- **Ancient Egypt:** Spices such as frankincense and myrrh were used in religious rituals and mummification processes, symbolizing purification and preservation.
- **Europe:** In medieval Europe, spices were used in religious ceremonies and to preserve the bodies of saints. They were also believed to have protective properties against evil spirits and diseases.

Modern Trends in Spice Consumption

In recent years, there has been a resurgence of interest in spices, driven by the growing popularity of global cuisines and the increasing awareness of their health benefits. The modern consumer is more adventurous, seeking out exotic flavors and incorporating spices into their daily diet.

- **Health and Wellness:** The health benefits of spices are being rediscovered, leading to their increased use in dietary supplements, functional foods, and natural remedies. Turmeric lattes, ginger shots, and cinnamon-infused beverages are examples of how spices are being integrated into health-conscious lifestyles.
- **Ethnic Cuisine:** The globalization of food culture has introduced people to a wide range of spices and cooking techniques. Ethnic restaurants and food products are popularizing spices like cumin, coriander, and chili, making them staples in many households.
- **Sustainability and Fair Trade:** Consumers are becoming more conscious of the environmental and social impacts of spice production. There is a growing demand for sustainably sourced and ethically traded spices, leading to the development of certification programs and fair trade initiatives.

The Future of the Spice Industry

The spice industry is poised for continued growth, driven by expanding global demand and innovative applications. Advances in agricultural practices, processing technologies, and supply chain management are enhancing the quality and availability of spices. Key trends shaping the future of the spice industry include:

- **Organic and Sustainable Farming:** There is an increasing focus on organic and sustainable farming practices to meet consumer demand for high-quality, environmentally friendly spices. Organic certification and sustainable sourcing are becoming important factors in the spice trade.
- **Innovation in Processing:** New technologies are being developed to improve the processing and preservation of spices, ensuring their flavor, aroma, and nutritional properties are retained. Cryogenic grinding, for instance, minimizes the loss of volatile oils and enhances the quality of ground spices.

» **Traceability and Transparency:** The spice industry is adopting digital technologies to enhance traceability and transparency in the supply chain. Blockchain technology, for example, is being used to track the journey of spices from farm to table, ensuring authenticity and quality.

» **Value Addition:** There is a growing trend towards value addition in the spice industry, with companies developing innovative products such as spice blends, extracts, and essential oils. These value-added products cater to diverse consumer needs and preferences.

1. Ginger (*Zingiber officinale* Rosc.)

Family: Zingiberaceae

Hindi: Adrak

Ginger is a perennial herbaceous plant, highly esteemed for its aromatic rhizome, which is extensively used as a spice and for medicinal applications. This versatile rhizome is a staple in various culinary traditions across the globe, utilized in forms such as fresh, dried, powdered, or processed into oil or juice. Its distinctive flavor and numerous therapeutic benefits have made it a favorite in the kitchen and in traditional medicine. Ginger is known for aiding digestion, alleviating nausea, and combating the flu and common cold. The bioactive compounds in ginger, particularly gingerol, possess significant anti-inflammatory and antioxidant properties, which contribute to its wide range of health benefits. In addition to its culinary and medicinal uses, ginger has been an integral part of cultural practices and traditional healing systems, particularly in Southeast Asia, where it originated. Its role in these cultures extends beyond food and medicine, influencing rituals and health practices passed down through generations. Ginger is also a critical component in various herbal formulations used to treat digestive disorders, respiratory issues, and inflammatory conditions.

Origin

The origin of ginger can be traced back to Southeast Asia, specifically to regions such as India and China, where it has been cultivated for thousands of years. Historical records and ancient texts highlight ginger's significant role in these regions' culinary and medicinal traditions. In India, ginger has been a crucial ingredient in Ayurvedic medicine, used to treat a variety of ailments ranging from digestive problems to inflammatory diseases. Similarly, in China, ginger has been a fundamental component of Traditional Chinese Medicine (TCM), valued for its warming properties and its ability to balance the body's internal energies.

The spread of ginger cultivation and its use expanded through trade routes to the Middle East, Europe, and eventually the Americas. During the medieval period, ginger was one of the most commonly traded spices, often mentioned in the same breath as black pepper. Its high demand and value made it a prized commodity in ancient trade networks, influencing the economies and cuisines of various regions. Today, ginger continues to be a globally significant spice, grown in tropical and subtropical regions worldwide.

Botany

Ginger is a herbaceous perennial plant characterized by its fleshy rhizome, which grows horizontally underground. The plant can reach up to 1 meter in height and has a pseudo-stem formed by the tightly wrapped leaf sheaths. The leaves are lanceolate, green, and arranged alternately along the stem. The inflorescence of ginger is a spike, which emerges directly from the rhizome. The flowers are typically yellowish-green with purple lips, adding to the plant's ornamental value.

The rhizome of ginger, which is the part used for culinary and medicinal purposes, develops underground and spreads horizontally. It is composed of several sections, each with a bud capable of sprouting into a new plant. The rhizomes are typically harvested when they are plump and have smooth, unwrinkled skin, usually 8-10 months after planting. This harvesting period ensures that the rhizomes have developed their full flavor and therapeutic properties. The botanical structure of ginger makes it well-suited for propagation and cultivation. The pseudo-stems, formed by the leaf sheaths, provide structural support and facilitate the plant's growth in dense clusters. The inflorescence spike, which bears the flowers, is directly connected to the rhizome, allowing for efficient nutrient transport and flower development. These botanical characteristics contribute to ginger's resilience and adaptability in various growing conditions.

Soil

Ginger thrives best in well-drained, friable loamy soils rich in organic matter. The ideal pH range for ginger cultivation is between 5.5 and 6.5. These soil conditions provide the necessary nutrients and structure for the optimal growth and development of the rhizomes. Loamy soils, characterized by their balanced mixture of sand, silt, and clay, offer excellent drainage and aeration, essential for preventing waterlogging and root diseases. While ginger can also be grown in sandy or clay loam soils, these soils must be well-drained. Sandy soils, though excellent for drainage, may require additional organic matter to enhance their nutrient content and water retention capacity. On the other hand, clay loam soils need to be well-aerated and managed to prevent compaction, which can hinder root growth and rhizome development. Heavy clay soils are generally unsuitable for ginger cultivation as they can retain excessive moisture, leading to waterlogging and rhizome rot. Alkaline soils are also not recommended as they can affect nutrient availability and uptake.

To ensure optimal soil conditions, it is essential to conduct soil tests and amend the soil as needed before planting ginger. Incorporating organic matter, such as compost or well-rotted manure, can improve soil structure, fertility, and moisture retention. Proper soil preparation, including ploughing and creating raised beds, further enhances the growing environment for ginger, promoting healthy and vigorous plant growth.

Land Preparation and Planting

Proper land preparation is a critical step in ensuring successful ginger cultivation. The soil must be ploughed 2-3 times to achieve a fine tilth, which creates a loose and friable texture conducive to rhizome expansion. Incorporating organic manure during this process enhances soil fertility and supports robust plant growth. Raised beds, typically 15 cm in height and 1 meter in width, are prepared with 40-50 cm spacing between them to facilitate irrigation and prevent waterlogging. Ginger rhizomes are cut into pieces weighing 20-25 grams, each containing at least one viable bud. These pieces are treated with fungicide to prevent fungal infections and promote healthy growth. The treated rhizome pieces are then planted at a spacing of 20-25 cm within rows and 25-30 cm between rows, with a planting depth of about 5-8 cm. The optimal time for planting ginger is during April-May, coinciding with the onset of the monsoon season. The increased moisture during this period supports the initial growth and establishment of the plants.

After planting, it is crucial to maintain consistent soil moisture to support germination and early growth. Light irrigation immediately after planting and regular watering during dry periods are essential to prevent the soil from drying out. Additionally, mulching the beds with organic materials such as straw or dried leaves helps conserve soil moisture, suppress weeds, and maintain a stable soil temperature, further promoting healthy plant growth.

Irrigation

Consistent and adequate irrigation is critical for ginger cultivation. Ginger requires frequent watering to maintain adequate soil moisture, especially during the early stages of growth and rhizome development. The first irrigation should be applied immediately after planting, followed by subsequent irrigations at intervals of 7-10 days during dry periods. Maintaining consistent soil moisture is essential for the proper growth and development of the rhizomes.

However, it is crucial to avoid waterlogging, as excessive moisture can lead to rhizome rot and other soil-borne diseases. Proper drainage systems should be in place to prevent water accumulation in the fields. Raised beds can help improve drainage, especially in areas with heavy rainfall or poorly drained soils. In addition to regular irrigation, it is beneficial to mulch the beds with organic materials such as straw or dried leaves. Mulching helps conserve soil moisture, reduce evaporation, suppress weeds, and maintain a stable soil temperature. These benefits contribute to healthy plant growth and higher yields.

Intercultural Operations

Intercultural operations such as weeding and earthing up are crucial in ginger cultivation. The first weeding is usually done 4-5 weeks after planting, followed by additional weedings at 3-4 week intervals. Weeds compete with ginger plants for nutrients, water, and light, which can significantly reduce crop yield and quality. Regular weeding helps maintain a clean and healthy growing environment.

Earthing up, performed during the second and third weedings, involves mounding soil around the base of the plants. This practice helps in covering the rhizomes properly, preventing them from being exposed to sunlight, which can cause damage. Earthing up also supports the plants and helps maintain a clean root zone, promoting healthy rhizome development. Other intercultural operations include removing diseased or damaged plants, thinning out crowded plants, and ensuring proper ventilation and light penetration. These practices help in maintaining a healthy crop and preventing the spread of pests and diseases.

Harvesting and Postharvest Management

Ginger is typically ready for harvest 8-10 months after planting when the leaves start to turn yellow and dry. The rhizomes are carefully lifted from the soil using a spade or digging fork, cleaned, and sun-dried for a day before marketing. For green ginger, which is harvested earlier (at 6-8 months), the rhizomes are more tender and are often used fresh or for making pickles.

After harvesting, the rhizomes are cleaned to remove soil and other debris. They are then washed thoroughly to ensure they are free from dirt and contaminants. Proper cleaning is essential for maintaining the quality and market value of the ginger. The cleaned rhizomes are then graded based on size, shape, and quality. Grading helps in categorizing ginger for different market uses, such as fresh consumption, processing, or export. Proper grading ensures uniformity and consistency in the final product. For dry ginger, the rhizomes are soaked in water overnight, scraped to remove the outer skin, and then sun-dried for about a week. The drying process reduces the moisture content, increasing the shelf life and preserving the quality of the ginger. Processed ginger should be stored in cool, dry conditions to prevent spoilage and maintain its aromatic and therapeutic properties.

Pests and Diseases

Pests

Ginger is susceptible to several pests that can affect its growth and yield:

- **Shoot Borer:** This pest causes dead hearts in the plant by boring into the shoots, leading to the wilting and death of affected shoots. Effective control measures include the use of pheromone traps, biological control agents, and timely application of insecticides.
- **Rhizome Fly:** The larvae of this fly feed on the rhizomes, causing significant damage and reducing the quality of the crop. Proper sanitation, use of traps, and chemical control can help manage this pest.
- **Nematodes:** These microscopic worms cause stunted growth and yellowing of leaves by attacking the root system. Crop rotation, organic amendments, and nematicides are effective control measures.

Diseases

Several diseases can affect ginger, impacting both the quality and quantity of the yield:

» **Rhizome Rot (Pythium spp.):** This disease causes soft rot of the rhizomes, leading to decay and loss of the crop. Good drainage, use of disease-free planting material, and application of fungicides can help control this disease.

» **Bacterial Wilt (Ralstonia solanacearum):** This bacterial infection results in yellowing and wilting of the plants, severely affecting their growth. Crop rotation, resistant varieties, and soil solarization are effective control measures.

» **Leaf Spot (Phyllosticta spp.):** This fungal disease causes brown spots on the leaves, reducing the photosynthetic ability of the plant and affecting overall health. Regular monitoring, proper sanitation, and fungicide application can manage this disease.

Control Measures

Effective management of pests and diseases is crucial for successful ginger cultivation. The following measures can help control these issues:

» **Use Disease-Free Planting Material:** Start with healthy, disease-free rhizomes to prevent the introduction of pests and diseases into the field.

» **Practice Crop Rotation:** Rotate ginger with non-host crops to break the life cycle of pests and pathogens.

» **Apply Recommended Fungicides and Pesticides:** Use appropriate chemical treatments as per recommended guidelines to control pest and disease outbreaks.

» **Maintain Proper Drainage:** Ensure that the soil is well-drained to avoid waterlogging, which can lead to rhizome rot.

» **Use Organic Mulches:** Apply organic mulches to conserve soil moisture, suppress weeds, and improve soil health.

Ginger Varieties and Their Characteristics

Variety	Developing Institution	Special Features
Nadia	ICAR - Indian Institute of Spices Research	High pungency, suitable for dry ginger and processing.
Maran	ICAR - Indian Institute of Spices Research	Large rhizomes, high yield, and moderately resistant to rhizome rot.
Rio-de-Janeiro	Kerala Agricultural University	High gingerol content, suitable for fresh consumption.
Varada	ICAR - Indian Institute of Spices Research	High yielding, tolerant to pests and diseases.

Himachal	CSK Himachal Pradesh Agricultural University	High yield, suitable for cultivation in North Indian hills.
Suprabha	ICAR - Indian Institute of Spices Research	High oil content, good for extraction and medicinal purposes.
Suruchi	Indian Institute of Spices Research	High yielding variety, resistant to major pests and diseases.
Suravi	ICAR - Indian Institute of Spices Research	Early maturing, suitable for fresh and dried ginger.
Karthika	Kerala Agricultural University	Good for fresh market and processing, high gingerol content.
Rejatha	Indian Institute of Spices Research	High yielding, aromatic, and resistant to rhizome rot.

Area and Production of Ginger

State	Area (hectares)	Production (tonnes)	Major Varieties Grown
Kerala	30,000	155,000	Rio-de-Janeiro, Maran, Varada
Assam	20,000	95,000	Nadia, Maran, Rio-de-Janeiro
Meghalaya	14,000	70,000	Nadia, Himachal, Varada
Mizoram	8,000	37,000	Nadia, Maran, Varada
Orissa	7,000	32,000	Suprabha, Suravi, Varada
Sikkim	5,000	22,000	Nadia, Himachal, Varada
West Bengal	5,000	25,000	Nadia, Maran, Rio-de-Janeiro
Karnataka	3,000	15,000	Rio-de-Janeiro, Varada, Suravi
Arunachal Pradesh	2,000	10,000	Nadia, Maran, Rio-de-Janeiro
Nagaland	1,000	5,000	Nadia, Varada, Himachal

Manuring and Fertilization

Ginger requires substantial amounts of both organic and inorganic fertilizers to achieve optimal growth. The application of farmyard manure at a rate of 20-25 tonnes per hectare is essential during land preparation. This organic matter enriches the soil, improving its structure, fertility, and water-holding capacity. In addition to organic manure, a balanced fertilizer regimen is crucial to provide the necessary nutrients for ginger growth:

» **Nitrogen (N):** 75-100 kg per hectare.

» **Phosphorus (P_2O_5):** 50-75 kg per hectare.

» **Potassium (K_2O):** 100-125 kg per hectare.

Half of the nitrogen and all of the phosphorus and potassium should be applied as a basal dose during the final land preparation. This initial fertilization ensures that the young plants have access to essential nutrients during the early stages of growth. The remaining nitrogen is applied in two split doses during the early growth stages and root development. This staged application supports continuous growth and helps in developing well-formed rhizomes.

It is also beneficial to apply micronutrients such as zinc, boron, and magnesium, which are essential for various physiological functions in the plant. Foliar sprays of these micronutrients can be applied during the growing season to ensure the plants receive adequate nutrition. Regular soil testing and monitoring of plant health can help in adjusting the fertilization regimen as needed, ensuring optimal growth and yield.

Ginger is a heavy feeder and responds well to organic and inorganic fertilizers. Here is the recommended fertilizer schedule:

Fertilizer	Quantity per Hectare	Time of Application
Farmyard Manure	25-30 tonnes	At the time of land preparation
Nitrogen (N)	75-100 kg	Half as basal dose, half in two splits at 60 and 120 days after planting
Phosphorus (P_2O_5)	50-75 kg	Entirely as basal dose
Potassium (K_2O)	75-100 kg	Entirely as basal dose

Micronutrients like Zinc and Boron can be applied if deficiencies are observed.

Table: Summary of Ginger Cultivation Practices

Aspect	Details
Climate	Warm and humid, 25-30°C temperature range, 150-300 cm annual rainfall.
Soil	Well-drained loamy soil, rich in organic matter, pH 5.5-6.5.
Land Preparation	Plough 2-3 times, add organic manure, prepare beds 15 cm height, 1 m width.
Planting Material	Rhizome pieces (20-25 g) with at least one viable bud, treated with fungicide.
Planting Time	April-May.
Spacing	20-25 cm within rows, 25-30 cm between rows.
Manures and Fertilizers	25-30 tonnes FYM, 75-100 kg N, 50-75 kg P_2O_5, 75-100 kg K_2O per hectare.
Irrigation	Frequent, at intervals of 7-10 days during dry periods. Avoid waterlogging.
Intercultural Operations	Weeding at 4-5 weeks, subsequent weedings at 3-4 week intervals, earthing up during second and third weeding.
Harvesting	8-10 months after planting for mature ginger, 6-8 months for green ginger.

Yield	15-25 tonnes per hectare.
Post-Harvest Management	Clean, wash, grade, dry or cure rhizomes, store in cool, dry conditions.
Pests	Shoot Borer, Rhizome Fly, Nematodes.
Diseases	Rhizome Rot (Pythium spp.), Bacterial Wilt (Ralstonia solanacearum), Leaf Spot (Phyllosticta spp.).
Control Measures	Use disease-free planting material, crop rotation, apply fungicides and pesticides, maintain drainage, use organic mulches.

2. Turmeric (*Curcuma longa* L.)

Family: Zingiberaceae

Hindi: Haldi

Turmeric is a perennial herbaceous plant renowned for its vibrant yellow rhizomes, which are extensively used as a spice, dye, and in traditional medicine. The rhizomes of turmeric contain bioactive compounds known as curcuminoids, with curcumin being the most prominent. Curcumin is celebrated for its potent anti-inflammatory, antioxidant, and antimicrobial properties, making turmeric a valuable ingredient in both culinary and medicinal contexts. In South Asian cuisine, turmeric is a staple, imparting a distinctive color and flavor to a variety of dishes. Beyond its culinary uses, turmeric is also employed in cosmetics and medicinal preparations, reflecting its versatility and cultural significance.

The therapeutic properties of turmeric have been recognized and utilized in traditional medicine systems for centuries. In Ayurveda, turmeric is used to treat a wide range of ailments, including digestive disorders, respiratory issues, and inflammatory conditions. Its antimicrobial properties make it effective against various infections, while its antioxidant effects help combat oxidative stress and promote overall health. The incorporation of turmeric in daily diets and traditional remedies underscores its importance in promoting health and well-being. Turmeric's vibrant color is attributed to the presence of curcuminoids, which also contribute to its health benefits. This color makes turmeric a popular natural dye, used in textiles, cosmetics, and food products. The versatility of turmeric extends to its use in rituals and cultural practices, particularly in South Asia, where it holds symbolic significance in various ceremonies and traditions.

Origin

Turmeric is believed to have originated in South Asia, particularly in India. It has been cultivated in the Indian subcontinent for thousands of years, playing a significant role in the region's culture, cuisine, and traditional medicine. Historical records and ancient texts highlight turmeric's extensive use in Ayurvedic and Chinese medicine, where it has been utilized for its healing properties and as a dietary staple. The long history of turmeric cultivation and use in these regions reflects its importance and widespread application in various cultural contexts.

The spread of turmeric cultivation and use expanded through trade routes to other parts of Asia, the Middle East, Africa, and eventually to the Americas. The spice trade played a crucial role in the dissemination of turmeric, making it a globally recognized and valued commodity. Today, turmeric is grown in tropical and subtropical regions worldwide, with India remaining the largest producer and exporter. Turmeric's journey from its origins in South Asia to its global presence today highlights its adaptability and the universal appeal of its culinary and medicinal benefits. Its enduring significance in traditional medicine systems and modern health practices underscores the timeless value of this remarkable plant.

Botany

Turmeric is a perennial herbaceous plant with a short stem and large, oblong leaves. The plant typically grows to a height of 1-1.5 meters, forming a dense clump of foliage that provides a lush, green backdrop in cultivated fields. The rhizome, the commercially important part of the plant, is tuberous and has a rough, segmented skin. Inside, the rhizomes are aromatic and display a characteristic yellow to orange color, which is the source of turmeric's distinctive hue.

The botanical structure of turmeric includes a pseudo-stem formed by the leaf sheaths, which supports the large, oblong leaves that arise alternately along the stem. The inflorescence of turmeric is a spike, which bears yellow flowers. These flowers are typically small and clustered, adding to the ornamental value of the plant. The flowering spike emerges directly from the rhizome, indicating the close connection between the reproductive and vegetative parts of the plant. Turmeric plants are propagated through rhizomes rather than seeds. This vegetative propagation method involves planting pieces of the rhizome, each containing at least one bud, which can sprout and develop into a new plant. This method of propagation ensures genetic uniformity and helps maintain the desirable traits of cultivated turmeric varieties. The ability to propagate through rhizomes also makes turmeric cultivation more manageable, as it allows farmers to select healthy, disease-free planting material.

The growth cycle of turmeric involves a period of active vegetative growth, followed by the development and maturation of the rhizomes. The plant thrives in warm, humid conditions with well-distributed rainfall, which supports vigorous growth and rhizome development. Proper soil management, including the incorporation of organic matter and maintaining optimal soil moisture, is crucial for successful turmeric cultivation. The harvested rhizomes undergo processing, which typically involves cleaning, boiling, drying, and polishing to produce the familiar turmeric powder used in culinary and medicinal applications. The processing methods can vary, but the goal is to preserve the bioactive compounds and ensure the rhizomes' quality and shelf life. The final product, whether in the form of whole rhizomes, slices, or powder, is valued for its flavor, color, and health benefits.

Varieties

Several varieties of turmeric are cultivated, each differing in yield, curcumin content, and suitability to specific climates.

Variety	Special Features
Alleppey Finger	High curcumin content, suitable for extraction and medicinal use.
Erode	Known for its bright yellow color and high yield.
Rajapuri	Large rhizomes, preferred for culinary use.
Sangli	High curcumin content, widely used in the spice industry.
Lakadong	High curcumin content, used in both culinary and medicinal preparations.
Prathibha	High yielding variety developed by IISR, Kerala.
Suvarna	Suitable for cultivation in South India, high yielding with moderate curcumin content.

Climate

Turmeric thrives in warm and humid climates, which are ideal for its growth and development. The plant requires a temperature range of 20-30°C, making it well-suited for tropical and subtropical regions. Turmeric also benefits from well-distributed annual rainfall, ideally between 150-300 cm, ensuring adequate moisture throughout its growth cycle. These climatic conditions promote vigorous growth and high yields.

Turmeric can be cultivated from sea level up to an altitude of 1500 meters, showcasing its adaptability to various elevations. However, the plant prefers partial shade, making it an excellent candidate for agroforestry systems where it can grow under the canopy of larger trees. This preference for shaded environments helps protect the delicate rhizomes from direct sunlight, which can cause damage and affect the quality of the yield.

In addition to temperature and rainfall, humidity plays a crucial role in turmeric cultivation. High humidity levels help maintain soil moisture, essential for the growth of the rhizomes. Areas with consistent humidity levels ensure that the plant receives the necessary moisture without the risk of waterlogging, which can be detrimental to turmeric cultivation. The combination of warm temperatures, adequate rainfall, and high humidity creates an ideal environment for turmeric to thrive. Furthermore, the timing of planting and harvesting is influenced by climatic conditions. Planting typically occurs with the onset of the monsoon season, providing the young plants with the necessary moisture for establishment. The monsoon rains help in settling the soil around the rhizomes, promoting healthy root development. Harvesting is usually done before the onset of winter or dry season to avoid exposing the rhizomes to adverse weather conditions that could affect their quality.

Area and Production of Turmeric

State	Area (hectares)	Production (tonnes)	Major Varieties Grown
Tamil Nadu	31,000	159,000	Erode, BSR-1, Suguna
Andhra Pradesh	28,500	164,000	Mydukur, Tekurpet, Armoor
Maharashtra	28,000	132,000	Rajapuri, Sangli, Salem
Odisha	22,000	52,000	Roma, Prathibha, Suvarna
Karnataka	19,000	58,000	Alleppey, Mydukur, Rajapuri
West Bengal	14,000	63,000	Suguna, Sona, Lakadong
Assam	10,000	43,000	Lakadong, Megha Turmeric-1, Sugandham
Kerala	8,000	25,000	Alleppey Finger, Prathibha, IISR Varieties
Gujarat	5,000	18,000	Rajapuri, Rani, Gauravi

Soil

Turmeric grows best in well-drained, fertile loamy soils rich in organic matter. The ideal pH range for turmeric cultivation is between 5.5 and 6.5. These soil conditions provide the necessary nutrients and structure for the optimal growth and development of the rhizomes. Loamy soils, characterized by their balanced mixture of sand, silt, and clay, offer excellent drainage and aeration, essential for preventing waterlogging and root diseases.

While turmeric can also be grown in sandy or clay loam soils, it is crucial that these soils are well-drained. Sandy soils, though excellent for drainage, may require additional organic matter to enhance their nutrient content and water retention capacity. On the other hand, clay loam soils need to be well-aerated and managed to prevent compaction, which can hinder root growth and rhizome development. Heavy clay soils are generally unsuitable for turmeric cultivation as they can retain excessive moisture, leading to waterlogging and rhizome rot. Alkaline soils are also not recommended as they can affect nutrient availability and uptake.

To ensure optimal soil conditions, it is essential to conduct soil tests and amend the soil as needed before planting turmeric. Incorporating organic matter, such as compost or well-rotted manure, can improve soil structure, fertility, and moisture retention. Proper soil preparation, including ploughing and creating raised beds, further enhances the growing environment for turmeric, promoting healthy and vigorous plant growth.

Land Preparation and Planting

Proper land preparation is a critical step in ensuring successful turmeric cultivation. The soil must be ploughed 2-3 times to achieve a fine tilth, which creates a loose and friable texture conducive to rhizome expansion. Incorporating organic manure during this process enhances soil fertility and supports robust plant growth. Raised beds or ridges and furrows are prepared to facilitate drainage and prevent waterlogging, which can be detrimental to turmeric rhizomes.

The planting material consists of rhizomes cut into pieces weighing 20-25 grams, each containing at least one viable bud. These pieces are treated with fungicide before planting to prevent fungal infections and promote healthy growth. The treated rhizome pieces are then planted at a spacing of 25-30 cm within rows and 45-60 cm between rows, with a planting depth of about 5-7 cm. The optimal time for planting turmeric is during April-May, coinciding with the onset of the monsoon season. The increased moisture during this period supports the initial growth and establishment of the plants.

After planting, it is crucial to maintain consistent soil moisture to support germination and early growth. Light irrigation immediately after planting and regular watering during dry periods are essential to prevent the soil from drying out. Additionally, mulching the beds with organic materials such as straw or dried leaves helps conserve soil moisture, suppress weeds, and maintain a stable soil temperature, further promoting healthy plant growth.

Manuring and Fertilization

Turmeric requires substantial amounts of both organic and inorganic fertilizers to achieve optimal growth. The application of farmyard manure at a rate of 20-25 tonnes per hectare is essential during land preparation. This organic matter enriches the soil, improving its structure, fertility, and water-holding capacity. In addition to organic manure, a balanced fertilizer regimen is crucial to provide the necessary nutrients for turmeric growth:

» **Nitrogen (N):** 75-100 kg per hectare.
» **Phosphorus (P_2O_5):** 50-75 kg per hectare.
» **Potassium (K_2O):** 100-125 kg per hectare.

Half of the nitrogen and all of the phosphorus and potassium should be applied as a basal dose during the final land preparation. This initial fertilization ensures that the young plants have access to essential nutrients during the early stages of growth. The remaining nitrogen is applied in two split doses during the early growth stages and root development. This staged application supports continuous growth and helps in developing well-formed rhizomes.

It is also beneficial to apply micronutrients such as zinc, boron, and magnesium, which are essential for various physiological functions in the plant. Foliar sprays of these micronutrients can be applied during the growing season to ensure the plants receive adequate nutrition. Regular soil testing and monitoring of plant health can help in adjusting the fertilization regimen as needed, ensuring optimal growth and yield.

Turmeric is a heavy feeder and responds well to organic and inorganic fertilizers. Here is the recommended fertilizer schedule:

Fertilizer	Quantity per Hectare	Time of Application
Farmyard Manure	25-30 tonnes	At the time of land preparation
Nitrogen (N)	75-100 kg	Half as basal dose, half in two splits at 60 and 120 days after planting
Phosphorus (P_2O_5)	50-75 kg	Entirely as basal dose
Potassium (K_2O)	75-100 kg	Entirely as basal dose

Micronutrients like Zinc and Boron can be applied if deficiencies are observed.

Turmeric Varieties and Their Characteristics

Variety	Developing Institution	Special Features
Alleppey Finger	Kerala Agricultural University	High curcumin content, suitable for extraction and medicinal use.
Erode	Tamil Nadu Agricultural University	Known for its bright yellow color and high yield.
Rajapuri	Gujarat Agricultural University	Large rhizomes, preferred for culinary use.
Sangli	Mahatma Phule Krishi Vidyapeeth	High curcumin content, widely used in the spice industry.
Lakadong	Indian Institute of Spices Research	High curcumin content, used in both culinary and medicinal preparations.
Prathibha	Indian Institute of Spices Research	High yielding variety, suitable for cultivation in South India.
Suvarna	Kerala Agricultural University	High yielding with moderate curcumin content, suitable for South India.
Sugandham	Indian Institute of Spices Research	High oil content, aromatic, and suitable for culinary use.
Roma	Indian Institute of Spices Research	High yielding, moderately resistant to leaf blotch.
Suranjana	Indian Institute of Spices Research	Early maturing variety with high yield.

Irrigation

Consistent and adequate irrigation is critical for turmeric cultivation. Turmeric requires frequent watering to maintain adequate soil moisture, especially during the early stages of growth and rhizome development. The first irrigation should be applied immediately after planting, followed by subsequent irrigations at intervals of 7-10 days during dry periods. Maintaining consistent soil moisture is essential for the proper growth and development of the rhizomes.

However, it is crucial to avoid waterlogging, as excessive moisture can lead to rhizome rot and other soil-borne diseases. Proper drainage systems should be in place to prevent water accumulation in the fields. Raised beds or ridges can help improve drainage, especially in areas with heavy rainfall or poorly drained soils. In addition to regular irrigation, it is beneficial to mulch the beds with organic materials such as straw or dried leaves. Mulching helps conserve soil moisture, reduce evaporation, suppress weeds, and maintain a stable soil temperature. These benefits contribute to healthy plant growth and higher yields.

Intercultural Operations

Intercultural operations such as weeding and earthing up are crucial in turmeric cultivation. The first weeding is usually done 4-5 weeks after planting, followed by additional weedings at 3-4 week intervals. Weeds compete with turmeric plants for nutrients, water, and light, which can significantly reduce crop yield and quality. Regular weeding helps maintain a clean and healthy growing environment. Earthing up, performed during the second and third weedings, involves mounding soil around the base of the plants. This practice helps in covering the rhizomes properly, preventing them from being exposed to sunlight, which can cause damage. Earthing up also supports the plants and helps maintain a clean root zone, promoting healthy rhizome development.

Other intercultural operations include removing diseased or damaged plants, thinning out crowded plants, and ensuring proper ventilation and light penetration. These practices help in maintaining a healthy crop and preventing the spread of pests and diseases.

Harvesting and Postharvest Management

Turmeric is typically ready for harvest 7-9 months after planting when the leaves start to turn yellow and dry. The rhizomes are carefully lifted from the soil using a spade or digging fork, cleaned, and sun-dried for a day before marketing. For green turmeric, which is harvested earlier (at 6-8 months), the rhizomes are more tender and are often used fresh or for making pickles.

After harvesting, the rhizomes are cleaned to remove soil and other debris. They are then washed thoroughly to ensure they are free from dirt and contaminants. Proper cleaning is essential for maintaining the quality and market value of the turmeric. The cleaned rhizomes are then boiled or steamed to gelatinize the starch and remove the raw odor. The rhizomes are then dried in the sun for 10-15 days, polished to remove the outer skin, and ground into powder. The drying process reduces the moisture content, increasing the shelf life and preserving the quality of the turmeric. Processed turmeric should be stored in airtight containers to retain its color and flavor.

Pests and Diseases

Pests

Turmeric is susceptible to several pests that can affect its growth and yield:

» **Shoot Borer:** This pest causes dead hearts in the plant by boring into the shoots, leading to the wilting and death of affected shoots. Effective control measures include the use of pheromone traps, biological control agents, and timely application of insecticides.

» **Rhizome Fly:** The larvae of this fly feed on the rhizomes, causing significant damage and reducing the quality of the crop. Proper sanitation, use of traps, and chemical control can help manage this pest.

» **Nematodes:** These microscopic worms cause stunted growth and yellowing of leaves by attacking the root system. Crop rotation, organic amendments, and nematicides are effective control measures.

Diseases

Several diseases can affect turmeric, impacting both the quality and quantity of the yield:

» **Rhizome Rot (Pythium spp.):** This disease causes soft rot of the rhizomes, leading to decay and loss of the crop. Good drainage, use of disease-free planting material, and application of fungicides can help control this disease.

» **Leaf Blotch (Taphrina maculans):** This fungal disease causes yellowing and drying of leaves, reducing the photosynthetic ability of the plant and affecting overall health. Regular monitoring, proper sanitation, and fungicide application can manage this disease.

» **Leaf Spot (Colletotrichum spp.):** This fungal disease causes brown spots on the leaves, reducing the photosynthetic ability of the plant and affecting overall health. Regular monitoring, proper sanitation, and fungicide application can manage this disease.

Control Measures

Effective management of pests and diseases is crucial for successful turmeric cultivation. The following measures can help control these issues:

» **Use Disease-Free Planting Material:** Start with healthy, disease-free rhizomes to prevent the introduction of pests and diseases into the field.

» **Practice Crop Rotation:** Rotate turmeric with non-host crops to break the life cycle of pests and pathogens.

» **Apply Recommended Fungicides and Pesticides:** Use appropriate chemical treatments as per recommended guidelines to control pest and disease outbreaks.

- **Maintain Proper Drainage:** Ensure that the soil is well-drained to avoid waterlogging, which can lead to rhizome rot.
- **Use Organic Mulches:** Apply organic mulches to conserve soil moisture, suppress weeds, and improve soil health.

Table: Summary of Turmeric Cultivation Practices

Aspect	Details
Climate	Warm and humid, 20-30°C temperature range, 150-300 cm annual rainfall.
Soil	Well-drained loamy soil, rich in organic matter, pH 5.5-6.5.
Land Preparation	Plough 2-3 times, add organic manure, prepare raised beds or ridges and furrows.
Planting Material	Rhizome pieces (20-25 g) with at least one viable bud, treated with fungicide.
Planting Time	April-May.
Spacing	25-30 cm within rows, 45-60 cm between rows.
Manures and Fertilizers	25-30 tonnes FYM, 75-100 kg N, 50-75 kg P_2O_5, 75-100 kg K_2O per hectare.
Irrigation	Frequent, at intervals of 7-10 days during dry periods. Avoid waterlogging.
Intercultural Operations	Weeding at 4-5 weeks, subsequent weedings at 3-4 week intervals, earthing up during second and third weeding.
Harvesting	7-9 months after planting for mature turmeric, 6-8 months for green turmeric.
Yield	15-25 tonnes per hectare.
Post-Harvest Management	Clean, wash, boil or steam, sun-dry, polish, grind into powder, store in airtight containers.
Pests	Shoot Borer, Rhizome Fly, Nematodes.
Diseases	Rhizome Rot (Pythium spp.), Leaf Blotch (Taphrina maculans), Leaf Spot (Colletotrichum spp.).
Control Measures	Use disease-free planting material, crop rotation, apply fungicides and pesticides, maintain drainage, use organic mulches.

Detailed Cultivation Practices

Land Preparation

- **Ploughing:** The land should be ploughed deeply 2-3 times to bring it to a fine tilth. This helps in the removal of weeds and incorporates organic matter into the soil.
- **Manuring:** Apply 25-30 tonnes of well-decomposed farmyard manure per hectare during the last ploughing. This improves soil fertility and structure.
- **Bed Preparation:** Raised beds of 15-20 cm height and 1-meter width are prepared with a spacing of 40-50 cm between the beds to facilitate drainage and ease of planting.

Planting

- **Selection of Rhizomes:** Choose healthy, disease-free rhizomes for planting. Cut the rhizomes into 4-5 cm long pieces, each weighing about 20-25 grams and having at least one viable bud.
- **Treatment of Rhizomes:** Treat the rhizome pieces with a fungicide solution (e.g., Mancozeb 3g/liter of water) to prevent fungal infections.
- **Planting Method:** Plant the treated rhizome pieces in the prepared beds at a depth of 5-7 cm. Maintain a spacing of 25-30 cm within rows and 45-60 cm between rows.
- **Mulching:** Mulch the beds with green leaves or organic mulch to conserve moisture, suppress weeds, and maintain soil temperature.

Nutrient Management

- **Basal Fertilization:** Apply half of the nitrogen (N) and the full dose of phosphorus (P_2O_5) and potassium (K_2O) as a basal dose before planting.
- **Top Dressing:** Apply the remaining nitrogen in two split doses at 60 and 120 days after planting. This ensures a steady supply of nutrients during the critical growth stages.
- **Micronutrient Management:** If micronutrient deficiencies are observed, apply appropriate foliar sprays of zinc, boron, or other deficient nutrients.

Irrigation

- **Initial Irrigation:** Give the first irrigation immediately after planting to settle the soil around the rhizomes.
- **Subsequent Irrigations:** Irrigate the crop at intervals of 7-10 days during dry periods. Ensure that the soil remains moist but not waterlogged.
- **Avoid Waterlogging:** Turmeric is sensitive to waterlogging, which can lead to rhizome rot. Ensure proper drainage to avoid standing water in the field.

Weed Management

- **Weeding:** Perform the first weeding 4-5 weeks after planting. Subsequent weedings are done at 3-4 week intervals. Weeding can be combined with earthing up to cover the exposed rhizomes.
- **Mulching:** Apply organic mulch to suppress weed growth and conserve soil moisture. Mulching also helps in maintaining soil temperature and adds organic matter to the soil as it decomposes.

Earthing Up

- **First Earthing Up:** The first earthing up is done during the second weeding, around 8-10 weeks after planting. This helps in covering the rhizomes and promoting better growth.
- **Second Earthing Up:** Perform the second earthing up during the third weeding, around 16-18 weeks after planting. This ensures that the rhizomes remain well-covered and protected.

Harvesting

- **Mature Turmeric:** Harvesting is done 7-9 months after planting when the leaves start turning yellow and dry. The rhizomes are lifted carefully with a spade or digging fork.
- **Green Turmeric:** For green turmeric, harvesting is done at 6-8 months. The rhizomes are harvested before they are fully mature and used fresh.

Post-Harvest Processing

- **Cleaning and Washing:** The harvested rhizomes are cleaned and washed thoroughly to remove soil and other impurities.
- **Boiling/Steaming:** The cleaned rhizomes are boiled or steamed for 30-45 minutes to gelatinize the starch and remove the raw odor.
- **Drying:** The boiled rhizomes are sun-dried for 10-15 days until they reach a moisture content of 8-10%. Proper drying ensures long shelf life and prevents mold growth.
- **Polishing:** The dried rhizomes are polished to remove the rough outer skin. This can be done manually or using mechanical polishers.
- **Grinding:** The polished rhizomes are ground into a fine powder, which is then sieved and packed in airtight containers to retain its color, flavor, and medicinal properties.

Storage

- **Storage Conditions:** Store turmeric powder in cool, dry, and dark conditions to preserve its quality. Airtight containers are recommended to prevent moisture absorption and contamination.
- **Shelf Life:** Properly processed and stored turmeric powder can have a shelf life of up to 2 years.

Pests and Diseases Management

- Shoot Borer: Use neem-based insecticides or pheromone traps to control shoot borers. Regular monitoring and early intervention are crucial.
- Rhizome Fly: Maintain proper field hygiene and use recommended insecticides to control rhizome fly infestation.

- » Nematodes: Practice crop rotation and use nematicides if necessary. Organic soil amendments can also help reduce nematode populations.
- » Rhizome Rot: Ensure proper drainage to prevent waterlogging. Use disease-free planting material and treat rhizomes with fungicides before planting.
- » Leaf Blotch and Leaf Spot: Apply fungicides like Mancozeb or Copper Oxychloride to control leaf diseases. Ensure good air circulation in the field by maintaining proper plant spacing.

3. Coriander (*Coriandrum sativum* L.)

Introduction

Coriander, commonly referred to as cilantro or Chinese parsley, is a herbaceous annual plant belonging to the family Apiaceae. This versatile herb is extensively cultivated for its leaves and seeds, both of which are integral to various culinary traditions worldwide. The leaves are known for their fresh, citrusy flavor, while the seeds are appreciated for their warm, nutty aroma and are a staple in spice blends across diverse cuisines. Coriander is an essential ingredient in both fresh and dried forms, used to enhance the flavor of a multitude of dishes.

The culinary and medicinal uses of coriander date back thousands of years, with historical records indicating its use in ancient Egyptian, Greek, and Roman cultures. Its significance extends beyond the kitchen, as coriander seeds are also used in traditional medicine to treat a variety of ailments, including digestive disorders and infections. The essential oil extracted from coriander seeds is employed in the fragrance industry and as a natural preservative.

Origin and Distribution

Coriander is believed to have originated in the Mediterranean region. It has been cultivated for millennia, spreading from its native range to various parts of the world through trade and exploration. Today, coriander is widely grown in countries such as India, Egypt, Morocco, China, and southern Europe. Each region has developed its unique varieties and cultivation practices, contributing to the global diversity of coriander.

In India, coriander is a significant spice crop, with the country being one of the largest producers and exporters of coriander seeds. The favorable climatic conditions and diverse agro-ecological zones in India support the extensive cultivation of coriander. Similarly, in Egypt and Morocco, coriander is grown on a large scale, with the seeds being a crucial component of local cuisine and traditional medicine. The widespread distribution of coriander reflects its adaptability to different climatic and soil conditions. It thrives in temperate and subtropical regions, making it a versatile crop suitable for cultivation in various parts of the world. The plant's ability to grow in diverse environments has contributed to its popularity and extensive use in global culinary and medicinal practices.

Botany

Coriander is a member of the Apiaceae family, which includes other well-known herbs and spices such as cumin, dill, and fennel. The plant is an herbaceous annual, characterized by its distinctive leaf morphology and aromatic seeds. The leaves are pinnate, with broad lobed base leaves and slender, feathery upper leaves, giving the plant a delicate and attractive appearance.

The flowers of coriander are small, white, or pale pink, and arranged in umbels, a characteristic inflorescence pattern of the Apiaceae family. These flowers eventually give way to small, globular seeds that are yellowish-brown and highly aromatic. The seeds contain essential oils that are rich in bioactive compounds, contributing to their flavor and medicinal properties. Coriander plants typically reach a height of 30-50 cm, with a branching habit that allows for dense foliage growth. The plant's life cycle is relatively short, with rapid germination and growth, making it an ideal crop for quick turnover in agricultural systems. The botanical characteristics of coriander, including its leaf and seed morphology, contribute to its versatility and wide range of uses in culinary and medicinal applications.

Climatic Requirements

Coriander thrives in temperate and subtropical regions with moderate rainfall and well-distributed precipitation. The ideal temperature range for coriander growth is between 15°C and 25°C. These temperatures are optimal for seed germination, vegetative growth, and seed production.

Coriander requires an annual rainfall of 600-1000 mm, with the water being well-distributed throughout the growing season. While coriander can tolerate short periods of drought, prolonged dry spells can adversely affect growth and yield. Conversely, excessive rainfall or waterlogging can lead to root diseases and poor seed quality. The plant prefers regions with moderate climates, avoiding extreme temperatures that can stress the plants and reduce productivity. Coriander can be grown at various altitudes, from sea level up to 1500 meters, showcasing its adaptability to different environmental conditions. The timing of planting is crucial, with sowing typically done in cooler months to avoid the adverse effects of high temperatures during the growing period.

Soil Requirements

Coriander grows best in well-drained, fertile loamy or sandy loam soils that are rich in organic matter and nutrients. The ideal soil pH for coriander cultivation ranges from 6.0 to 8.0. These soil conditions provide the necessary nutrients and structure for optimal growth and development of the plants.

Loamy soils, characterized by their balanced mixture of sand, silt, and clay, offer excellent drainage and aeration, essential for preventing waterlogging and root diseases. Sandy loam soils, with their higher sand content, also provide good drainage and are suitable for coriander cultivation. The addition of organic matter, such as compost or well-rotted manure, improves soil fertility and moisture retention, promoting healthy plant growth. Heavy clay soils and waterlogged conditions are detrimental to coriander cultivation, as they can lead to poor root development and increased susceptibility to diseases. Proper soil management, including regular soil testing and amendments, is essential to maintain the ideal soil conditions for coriander growth. Ensuring adequate drainage and avoiding waterlogging are critical to prevent root rot and other soil-borne diseases.

Land Preparation and Planting

Proper land preparation is a critical step in ensuring successful coriander cultivation. The soil should be ploughed deeply to create a fine tilth, which provides a loose and friable texture conducive to root development. Incorporating organic manure during land preparation enhances soil fertility and supports robust plant growth.

Sowing time varies depending on the region and climate:

» **Plains:** October to November.

» **Hills:** February to March.

The seeds are sown at a rate of 10-15 kg per hectare, either by broadcasting or drilling in rows. The recommended spacing is 30-45 cm between rows and 10-15 cm within rows, allowing adequate space for plant growth and air circulation. Thinning of seedlings is necessary to maintain proper spacing and reduce competition for nutrients and water.

The depth of planting should be about 1.5-2.0 cm, ensuring that the seeds are adequately covered with soil. After sowing, it is essential to provide light irrigation to promote seed germination and early seedling establishment. Maintaining soil moisture during the initial stages is crucial for uniform germination and healthy plant growth.

Manuring and Fertilization

Coriander requires a balanced supply of nutrients for optimal growth and yield. The application of farmyard manure at a rate of 10-15 tonnes per hectare during land preparation significantly improves soil fertility and structure. In addition to organic manure, a balanced fertilization regimen is essential:

» **Nitrogen (N):** 60 kg per hectare.

» **Phosphorus (P_2O_5):** 40 kg per hectare.

» **Potassium (K_2O):** 20 kg per hectare.

Half of the nitrogen and the full dose of phosphorus and potassium should be applied as a basal dose during sowing. The remaining nitrogen is applied as a top dressing after 30 days of sowing, supporting continuous growth and development. This staged application ensures that the plants receive the necessary nutrients during critical growth phases.

Regular soil testing and monitoring of plant health can help in adjusting the fertilization regimen as needed. The application of micronutrients such as zinc and boron, through foliar sprays, can further enhance plant growth and yield. Proper nutrient management is essential to maximize productivity and produce high-quality coriander leaves and seeds.

Irrigation

Coriander requires regular irrigation to maintain adequate soil moisture for optimal growth. The first irrigation should be applied immediately after sowing to promote seed germination and seedling establishment. Subsequent irrigations are provided at intervals of 10-15 days, depending on weather conditions and soil moisture levels.

It is crucial to avoid waterlogging, as excessive moisture can lead to root diseases and poor seed quality. Proper drainage systems should be in place to prevent water accumulation in the fields. In areas with heavy rainfall or poorly drained soils, raised beds or ridges can help improve drainage and prevent waterlogging. In addition to regular irrigation, mulching the beds with organic materials such as straw or dried leaves helps conserve soil moisture, reduce evaporation, suppress weeds, and maintain a stable soil temperature. These benefits contribute to healthy plant growth and higher yields.

Intercultural Operations

Intercultural operations such as weeding and thinning are crucial in coriander cultivation. The first weeding is usually done 4-5 weeks after sowing, followed by additional weedings at 3-4 week intervals. Weeds compete with coriander plants for nutrients, water, and light, which can significantly reduce crop yield and quality. Regular weeding helps maintain a clean and healthy growing environment. Thinning of seedlings is necessary to maintain proper spacing and reduce competition for nutrients and water. The recommended spacing is 10-15 cm between plants, allowing adequate space for growth and air circulation. Proper thinning ensures uniform plant development and maximizes yield potential. Other intercultural operations include removing diseased or damaged plants, thinning out crowded plants, and ensuring proper ventilation and light penetration. These practices help in maintaining a healthy crop and preventing the spread of pests and diseases.

Harvesting and Postharvest Management

Coriander leaves are typically harvested when the plants reach a height of 30-40 cm. The leaves are cut or plucked, and the plants are allowed to regrow for subsequent harvests. Regular harvesting of leaves promotes continuous growth and ensures a steady supply of fresh coriander.

For seed production, coriander is harvested when the seeds turn brown and start to dry. The entire plant is cut and left to dry in the field for a few days before threshing. The seeds are then separated from the plant, cleaned, and dried further in the sun for 4-5 days. Proper drying and storage of seeds are essential to maintain their quality and viability. The dried seeds should be stored in cool, dry, and airtight conditions to prevent moisture absorption and spoilage. Proper postharvest management ensures that the seeds retain their flavor, aroma, and germination potential.

Post-Harvest

» **Drying:** Sun-dry seeds for 4-5 days.

» **Threshing:** Separate seeds from the plant.

» **Storage:** Cool, dry, and airtight conditions.

Pests and Diseases

Pests

Coriander is susceptible to several pests that can affect its growth and yield:

» **Shoot Borer:** This pest causes dead hearts in the plant by boring into the shoots, leading to the wilting and death of affected shoots. Effective control measures include the use of pheromone traps, biological control agents, and timely application of insecticides.

» **Rhizome Fly:** The larvae of this fly feed on the rhizomes, causing significant damage and reducing the quality of the crop. Proper sanitation, use of traps, and chemical control can help manage this pest.

» **Nematodes:** These microscopic worms cause stunted growth and yellowing of leaves by attacking the root system. Crop rotation, organic amendments, and nematicides are effective control measures.

Diseases

Several diseases can affect coriander, impacting both the quality and quantity of the yield:

» **Rhizome Rot (Pythium spp.):** This disease causes soft rot of the rhizomes, leading to decay and loss of the crop. Good drainage, use of disease-free planting material, and application of fungicides can help control this disease.

» **Leaf Blotch (Taphrina maculans):** This fungal disease causes yellowing and drying of leaves, reducing the photosynthetic ability of the plant and affecting overall health. Regular monitoring, proper sanitation, and fungicide application can manage this disease.

» **Leaf Spot (Colletotrichum spp.):** This fungal disease causes brown spots on the leaves, reducing the photosynthetic ability of the plant and affecting overall health. Regular monitoring, proper sanitation, and fungicide application can manage this disease.

Control Measures

Effective management of pests and diseases is crucial for successful coriander cultivation. The following measures can help control these issues:

- **Use Disease-Free Planting Material:** Start with healthy, disease-free seeds to prevent the introduction of pests and diseases into the field.
- **Practice Crop Rotation:** Rotate coriander with non-host crops to break the life cycle of pests and pathogens.
- **Apply Recommended Fungicides and Pesticides:** Use appropriate chemical treatments as per recommended guidelines to control pest and disease outbreaks.
- **Maintain Proper Drainage:** Ensure that the soil is well-drained to avoid waterlogging, which can lead to root diseases.
- **Use Organic Mulches:** Apply organic mulches to conserve soil moisture, suppress weeds, and improve soil health.

Varieties and Their Characteristics

Variety	Developing Institution	Special Features
Coriander Co 1	Tamil Nadu Agricultural University	High yield, suitable for green leaf production.
Coriander Co 2	Tamil Nadu Agricultural University	High seed yield, resistant to pests and diseases.
Pant Haritama	GB Pant University of Agriculture and Technology	High yield, suitable for both green leaves and seeds.
Gujarat Coriander 1	Anand Agricultural University	High seed yield, early maturing.
Sindhu	Mahatma Phule Krishi Vidyapeeth	High yield, aromatic seeds.
Coriander Sadhana	Indian Institute of Spices Research	High yield, suitable for seed production, moderately resistant to powdery mildew.
CS-6	Chaudhary Charan Singh Haryana Agricultural University	High seed yield, suitable for rainfed conditions.
CS-11	Chaudhary Charan Singh Haryana Agricultural University	High yield, moderately resistant to stem gall disease.
Coriander Kalmi	Kerala Agricultural University	High yield, suitable for green leaves and seeds, tolerant to wilt.
Rajendra Swathi	Rajendra Agricultural University	High yield, aromatic seeds, resistant to wilt.

Area and Production of Coriander

State	Area (hectares)	Production (tonnes)	Major Varieties Grown
Rajasthan	120,000	125,000	CS-6, CS-11, Sindhu
Madhya Pradesh	90,000	110,000	Pant Haritama, Gujarat Coriander 1
Gujarat	50,000	55,000	Gujarat Coriander 1, CS-6
Andhra Pradesh	40,000	45,000	Rajendra Swathi, Coriander Kalmi
Tamil Nadu	30,000	35,000	Coriander Co 1, Coriander Co 2
Karnataka	25,000	30,000	CS-11, Sindhu
Uttar Pradesh	20,000	25,000	Pant Haritama, Gujarat Coriander 1
Maharashtra	15,000	20,000	Sindhu, Coriander Sadhana
Haryana	10,000	12,000	CS-6, CS-11
West Bengal	8,000	10,000	Rajendra Swathi, Coriander Kalmi

Uses

» **Culinary Uses:** Both leaves and seeds are used as spices in various cuisines. Leaves are used fresh, while seeds are used whole or ground.

» **Medicinal Uses:** Known for digestive, carminative, and anti-inflammatory properties. Used in traditional medicine for treating digestive issues and as an ingredient in various herbal remedies.

4. Cumin (*Cuminum cyminum* L.)

Introduction

Cumin is a flowering plant belonging to the family Apiaceae. Native to the region encompassing the Middle East and stretching east to India, cumin is highly prized for its seeds. These seeds are integral to the cuisines of many cultures around the world, used both whole and ground to impart a distinctive, warm flavor to a variety of dishes. Cumin's culinary significance is matched by its medicinal properties, as it has been used traditionally to treat digestive issues and other ailments.

Origin and Distribution

Cumin is believed to have originated in the Mediterranean region. It has a long history of cultivation, dating back to ancient times. The plant has spread far beyond its native range, now extensively grown in India, China, Indonesia, Iran, Turkey, Egypt, and several Mediterranean countries. Each of these regions has developed unique cultivation practices suited to their specific climates and soil conditions.

In India, cumin is particularly significant, as the country is one of the largest producers and consumers of cumin seeds. The favorable climatic conditions in the arid and semi-arid regions of India support extensive cumin cultivation. Similarly, in the Middle East and

Mediterranean regions, cumin is a staple spice, essential to local cuisines and traditional medicine practices. The widespread cultivation of cumin across different continents highlights its adaptability and importance in global agriculture. Its seeds are not only used in cooking but also valued for their therapeutic properties, including their ability to aid digestion, boost the immune system, and provide antioxidant benefits.

Botany

Cumin is an annual herb that exhibits distinct botanical characteristics:

- **Family:** Apiaceae
- **Plant Type:** Annual herb
- **Leaves:** Pinnate or bipinnate, with thread-like leaflets, giving the plant a delicate and feathery appearance.
- **Flowers:** Small, white or pink, arranged in umbels, a typical inflorescence of the Apiaceae family.
- **Seeds:** Small, elongated, ridged, and brownish, with a distinctive aroma that is both earthy and spicy.

The plant typically grows to a height of 30-50 cm, forming a dense foliage that supports its small, aromatic seeds. The seeds are highly valued for their flavor and aroma, which are due to the presence of essential oils and bioactive compounds. These botanical features make cumin a versatile and widely used spice in various culinary traditions.

Cumin plants have a relatively short growing season, with a life cycle that completes within 100-120 days. This rapid growth and maturation make cumin an ideal crop for farmers seeking quick returns. The plant's adaptability to different environmental conditions further enhances its suitability for diverse agricultural systems.

Climatic Requirements

Cumin thrives in dry regions with moderate rainfall and high sunlight exposure. The ideal temperature range for cumin cultivation is between 20°C and 30°C. These temperatures are conducive to seed germination, vegetative growth, and seed production.

Cumin requires an annual rainfall of 200-300 mm, which is sufficient to meet the plant's moisture needs without causing waterlogging. The plant is well-suited to regions with low humidity and high sunlight, as these conditions help reduce the risk of fungal diseases and promote healthy growth.

Cumin is typically sown in the cooler months to avoid the adverse effects of high temperatures during the growing season. The timing of planting is crucial, with sowing generally done in October to November. This ensures that the plants receive adequate moisture during the initial stages of growth and mature under optimal climatic conditions.

The plant's preference for dry and arid climates makes it suitable for cultivation in regions where other crops may struggle. Its ability to withstand high temperatures and low rainfall enhances its resilience and adaptability, making cumin a valuable crop for farmers in challenging environments.

Soil Requirements

Cumin grows best in well-drained sandy loam or loamy soils that are moderately fertile and rich in organic matter. The ideal soil pH for cumin cultivation ranges from 6.8 to 8.3. These soil conditions provide the necessary nutrients and structure for optimal growth and development of the plants.

Well-drained soils are essential for cumin cultivation, as the plant is susceptible to root diseases in waterlogged conditions. Sandy loam soils, characterized by their good drainage and aeration, are particularly suitable for cumin. The addition of organic matter, such as compost or well-rotted manure, improves soil fertility and moisture retention, promoting healthy plant growth. In areas with heavy clay soils, proper soil management practices, including deep ploughing and the creation of raised beds, are necessary to improve drainage and prevent waterlogging. Ensuring that the soil is well-drained and fertile is crucial for maximizing cumin yield and quality. Regular soil testing and amendments can help maintain the ideal soil conditions for cumin cultivation. The application of organic and inorganic fertilizers enhances soil fertility, supporting vigorous growth and high seed production.

Land Preparation and Planting

Proper land preparation is essential for successful cumin cultivation. The soil should be ploughed deeply to create a fine tilth, which provides a loose and friable texture conducive to root development. Incorporating organic manure during land preparation enhances soil fertility and supports robust plant growth. The optimal time for sowing cumin is from October to November. The seeds are sown at a rate of 12-15 kg per hectare, either by broadcasting or drilling in rows. The recommended spacing is 30-45 cm between rows and 15-20 cm within rows, allowing adequate space for plant growth and air circulation. Thinning of seedlings is necessary to maintain proper spacing and reduce competition for nutrients and water. The depth of planting should be about 1.5-2.0 cm, ensuring that the seeds are adequately covered with soil. After sowing, it is essential to provide light irrigation to promote seed germination and early seedling establishment. Maintaining soil moisture during the initial stages is crucial for uniform germination and healthy plant growth.

Manuring and Fertilization

Cumin requires a balanced supply of nutrients for optimal growth and yield. The application of farmyard manure at a rate of 10-15 tonnes per hectare during land preparation significantly improves soil fertility and structure. In addition to organic manure, a balanced fertilization regimen is essential:

- **Nitrogen (N):** 20 kg per hectare.
- **Phosphorus (P_2O_5):** 10 kg per hectare.
- **Potassium (K_2O):** 10 kg per hectare.

Half of the nitrogen and the full dose of phosphorus and potassium should be applied as a basal dose during sowing. The remaining nitrogen is applied as a top dressing after 30 days of sowing, supporting continuous growth and development. This staged application ensures that the plants receive the necessary nutrients during critical growth phases. Regular soil testing and monitoring of plant health can help in adjusting the fertilization regimen as needed. The application of micronutrients such as zinc and boron, through foliar sprays, can further enhance plant growth and yield. Proper nutrient management is essential to maximize productivity and produce high-quality cumin seeds.

Irrigation

Cumin requires light but regular irrigation to maintain adequate soil moisture for optimal growth. The first irrigation should be applied immediately after sowing to promote seed germination and seedling establishment. Subsequent irrigations are provided at critical stages, including at 30 days after sowing and during flowering and seed formation. It is crucial to avoid waterlogging, as excessive moisture can lead to root diseases and poor seed quality. Proper drainage systems should be in place to prevent water accumulation in the fields. In areas with heavy rainfall or poorly drained soils, raised beds or ridges can help improve drainage and prevent waterlogging. In addition to regular irrigation, mulching the beds with organic materials such as straw or dried leaves helps conserve soil moisture, reduce evaporation, suppress weeds, and maintain a stable soil temperature. These benefits contribute to healthy plant growth and higher yields.

Intercultural Operations

Intercultural operations such as weeding and thinning are crucial in cumin cultivation. The first weeding is usually done 4-5 weeks after sowing, followed by additional weedings at 3-4 week intervals. Weeds compete with cumin plants for nutrients, water, and light, which can significantly reduce crop yield and quality. Regular weeding helps maintain a clean and healthy growing environment. Thinning of seedlings is necessary to maintain proper spacing and reduce competition for nutrients and water. The recommended spacing is 15-20 cm between plants, allowing adequate space for growth and air circulation. Proper thinning ensures uniform plant development and maximizes yield potential. Other intercultural operations include removing diseased or damaged plants, thinning out crowded plants, and ensuring proper ventilation and light penetration. These practices help in maintaining a healthy crop and preventing the spread of pests and diseases.

Harvesting and Postharvest Management

Cumin is typically ready for harvest 100-120 days after sowing, when the seeds turn brown and are fully matured. The plants are usually harvested by hand, cutting the entire plant and allowing it to dry in the field for a few days. Once the plants are dry, they are threshed to separate the seeds from the plant material. Proper drying of the seeds is essential to reduce moisture content and prevent spoilage. The seeds should be spread out in a thin layer and dried in the sun for 4-5 days. After drying, the seeds are cleaned to remove any debris and impurities, ensuring high-quality cumin seeds. The cleaned seeds should be stored in cool, dry, and airtight conditions to maintain their flavor, aroma, and shelf life. Proper storage is crucial to prevent the seeds from absorbing moisture and becoming susceptible to mold and pests. Using moisture-proof containers and storing them in a low-humidity environment helps preserve the quality of the cumin seeds.

Pests and Diseases

Pests

Cumin is susceptible to several pests that can affect its growth and yield:

- **Shoot Borer:** This pest causes dead hearts in the plant by boring into the shoots, leading to the wilting and death of affected shoots. Effective control measures include the use of pheromone traps, biological control agents, and timely application of insecticides.
- **Rhizome Fly:** The larvae of this fly feed on the rhizomes, causing significant damage and reducing the quality of the crop. Proper sanitation, use of traps, and chemical control can help manage this pest.
- **Nematodes:** These microscopic worms cause stunted growth and yellowing of leaves by attacking the root system. Crop rotation, organic amendments, and nematicides are effective control measures.

Diseases

Several diseases can affect cumin, impacting both the quality and quantity of the yield:

- **Rhizome Rot (Pythium spp.):** This disease causes soft rot of the rhizomes, leading to decay and loss of the crop. Good drainage, use of disease-free planting material, and application of fungicides can help control this disease.
- **Leaf Blotch (Taphrina maculans):** This fungal disease causes yellowing and drying of leaves, reducing the photosynthetic ability of the plant and affecting overall health. Regular monitoring, proper sanitation, and fungicide application can manage this disease.
- **Leaf Spot (Colletotrichum spp.):** This fungal disease causes brown spots on the leaves, reducing the photosynthetic ability of the plant and affecting overall health. Regular monitoring, proper sanitation, and fungicide application can manage this disease.

Control Measures

Effective management of pests and diseases is crucial for successful cumin cultivation. The following measures can help control these issues:

- **Use Disease-Free Planting Material:** Start with healthy, disease-free seeds to prevent the introduction of pests and diseases into the field.
- **Practice Crop Rotation:** Rotate cumin with non-host crops to break the life cycle of pests and pathogens.
- **Apply Recommended Fungicides and Pesticides:** Use appropriate chemical treatments as per recommended guidelines to control pest and disease outbreaks.
- **Maintain Proper Drainage:** Ensure that the soil is well-drained to avoid waterlogging, which can lead to root diseases.
- **Use Organic Mulches:** Apply organic mulches to conserve soil moisture, suppress weeds, and improve soil health.

Varieties and Their Characteristics

Variety	Developing Institution	Special Features
RZ-19	ICAR-NRCSS, Ajmer	High yield, resistant to wilt and blight, suitable for Rajasthan.
RZ-209	ICAR-NRCSS, Ajmer	High yield, good seed quality, resistant to wilt.
GC-1	Gujarat Agricultural University	High yield, resistant to wilt, bold seed.
GC-2	Gujarat Agricultural University	High yield, good for rainfed conditions.
UC-198	University of Agricultural Sciences, Dharwad	High yield, suitable for Karnataka.
S-404	Haryana Agricultural University, Hisar	High yield, resistant to wilt and powdery mildew.
Hiranmayi	Kerala Agricultural University	High yield, tolerant to pests and diseases.
CO-1	Tamil Nadu Agricultural University	Early maturing, high yield, good seed quality.

Area and Production of Cumin

State	Area (hectares)	Production (tonnes)	Major Varieties Grown
Rajasthan	450,000	300,000	RZ-19, RZ-209
Gujarat	200,000	140,000	GC-1, GC-2
Haryana	20,000	12,000	S-404
Karnataka	15,000	10,000	UC-198
Madhya Pradesh	10,000	8,000	Local varieties
Tamil Nadu	5,000	4,000	CO-1

Uttar Pradesh	5,000	3,000	Local varieties
Kerala	3,000	2,000	Hiranmayi

Cumin (*Cuminum cyminum* L.): Uses

Culinary Uses

Cumin is an essential spice in many global cuisines, known for its warm, earthy flavor and distinctive aroma. It is used in both whole and ground forms and is a key ingredient in spice blends and seasonings such as curry powder, garam masala, chili powder, and adobo. Here are some of its prominent culinary uses:

1. **Flavoring Agent:** Cumin seeds are used to enhance the flavor of a variety of dishes, including soups, stews, sauces, and rice dishes. They are often dry roasted or fried in oil to release their aromatic oils.
2. **Spice Blends:** Ground cumin is a vital component of many spice blends and rubs, used to season meats, vegetables, and legumes.
3. **Baked Goods:** In some cultures, cumin seeds are added to bread and pastry doughs to impart a unique taste.
4. **Pickling:** Cumin seeds are used in pickling recipes, adding a distinctive flavor to preserved vegetables.
5. **Beverages:** Cumin is also used in certain traditional beverages, such as "jaljeera," a popular Indian summer drink made from cumin, mint, lemon, and other spices.

Medicinal Uses

Cumin has been valued for its medicinal properties for centuries, particularly in traditional medicine systems like Ayurveda and Traditional Chinese Medicine. It is known for its digestive, carminative, and anti-inflammatory properties. Key medicinal uses include:

1. **Digestive Aid:** Cumin is widely used to relieve indigestion, bloating, and gas. It stimulates the secretion of digestive enzymes and helps in the assimilation of nutrients.
2. **Anti-inflammatory:** The active compounds in cumin, such as thymoquinone, have anti-inflammatory effects that can help reduce pain and inflammation in conditions like arthritis.
3. **Antioxidant Properties:** Cumin seeds are rich in antioxidants, which help protect the body from oxidative stress and reduce the risk of chronic diseases.
4. **Antimicrobial:** Cumin has natural antimicrobial properties, making it useful in treating infections and promoting overall health.
5. **Respiratory Health:** Cumin is used to alleviate symptoms of respiratory conditions such as asthma and bronchitis by acting as a bronchodilator and expectorant.

Cosmetic Uses

Cumin is also utilized in the cosmetic industry for its beneficial effects on the skin and hair:

1. **Skin Care:** Cumin oil is used in skincare products for its antibacterial and anti-inflammatory properties. It helps treat acne, boils, and skin infections.
2. **Anti-Aging:** The antioxidants in cumin help fight free radicals, reducing signs of aging such as wrinkles and fine lines.
3. **Hair Care:** Cumin oil is used in hair care formulations to strengthen hair, prevent hair loss, and treat dandruff and scalp infections.

Industrial Uses

Beyond culinary and medicinal applications, cumin finds its way into various industrial uses:

1. **Flavoring Agents:** Cumin extracts and essential oils are used in the food industry to flavor sauces, marinades, and processed foods.
2. **Fragrance Industry:** Cumin oil is used in perfumes and aromatic products for its warm, spicy scent.
3. **Natural Preservative:** Due to its antimicrobial properties, cumin is used as a natural preservative in food products to extend shelf life.

Cultural and Ritual Uses

Cumin has cultural significance in many regions, where it is used in traditional rituals and ceremonies:

1. **Religious Offerings:** In some cultures, cumin is used in religious rituals and offerings due to its sacred status.
2. **Traditional Remedies**: Folk medicine practices often include cumin as a remedy for various ailments, leveraging its therapeutic properties passed down through generations.

5. Fennel (*Foeniculum vulgare* Mill.)

Introduction

Fennel, a member of the Apiaceae family, is a perennial herb characterized by its yellow flowers and feathery leaves. Highly aromatic and flavorful, fennel is utilized both as a spice and for medicinal purposes. It originates from the Mediterranean region, where it has been cultivated since ancient times. Today, fennel is grown worldwide, particularly in regions that offer the right climatic conditions. Fennel's aromatic seeds are integral to various culinary traditions, providing a sweet, anise-like flavor to dishes, while its medicinal properties make it valuable in traditional remedies.

Fennel's diverse applications extend beyond culinary uses. Its essential oils are extracted for use in perfumes and aromatherapy, while the seeds and leaves are used in herbal teas and health supplements. The plant's aesthetic appeal also makes it a popular choice in ornamental gardening.

Origin and Distribution

Fennel is believed to have originated in the Mediterranean region, where it has been an essential part of local diets and traditional medicine for centuries. Historical texts and archaeological findings suggest that ancient Egyptians, Greeks, and Romans highly valued fennel for its culinary and medicinal uses. The plant's significance extended beyond food and medicine; it was also associated with cultural rituals and beliefs.

Over time, fennel cultivation spread across Europe, the Middle East, India, China, and other regions with suitable climates. Each region has developed unique cultivation practices tailored to local conditions. For instance, in India, fennel is a staple spice used in various traditional dishes and Ayurvedic medicine. Similarly, in Italy and France, fennel bulbs are widely used in cooking, reflecting the plant's versatility and global appeal. Today, fennel is grown in numerous countries, each contributing to the global supply of fennel seeds, essential oils, and other fennel-based products. The plant's adaptability to different climates and soil types has facilitated its widespread cultivation, making it a valuable crop in many agricultural systems.

Botany

Fennel is a perennial herb that can also be grown as an annual. It belongs to the Apiaceae family, which includes other well-known herbs and spices such as parsley, dill, and cumin. Key botanical characteristics of fennel include:

- **Family:** Apiaceae
- **Plant Type:** Perennial herb, often cultivated as an annual for commercial purposes.
- **Leaves:** Finely dissected and feathery, giving the plant a delicate appearance. The leaves are bright green and have a pleasant aroma.
- **Flowers:** Small, yellow, and arranged in large, flat-topped umbels. These flower clusters are not only attractive but also crucial for seed production.
- **Seeds:** Oval, ridged, and greenish-brown with a sweet, anise-like flavor. The seeds are highly aromatic and contain essential oils that contribute to their distinctive taste and medicinal properties.

Fennel plants can grow up to 2 meters in height, forming a tall, bushy foliage. The feathery leaves and bright yellow flowers add aesthetic value, making fennel a popular choice for both culinary and ornamental gardening. The plant's robust growth habit and adaptability make it a versatile crop suitable for various agricultural and horticultural practices.

The botanical structure of fennel includes a deep taproot that helps the plant access nutrients and water from deeper soil layers. This root system supports the plant's growth in different soil types, contributing to its resilience and productivity.

Climatic Requirements

Fennel thrives in mild, temperate climates with well-distributed rainfall and plenty of sunlight. The ideal temperature range for fennel cultivation is between 15°C and 25°C. These temperatures are conducive to seed germination, vegetative growth, and seed production.

Fennel requires an annual rainfall of 500-600 mm, which provides sufficient moisture for healthy plant growth without causing waterlogging. The plant prefers regions with moderate climates, avoiding extreme temperatures that can stress the plants and reduce productivity. Fennel is particularly well-suited to areas with mild winters and cool summers, where the growing conditions support optimal development. The timing of planting is crucial for fennel cultivation. In regions with a Mediterranean climate, sowing typically occurs in the fall, allowing the plants to establish before the onset of winter. In other temperate regions, planting can be timed to avoid the hottest parts of the summer, ensuring that the plants grow under optimal conditions. Fennel's adaptability to different climatic conditions makes it a versatile crop. However, it performs best in regions that offer a combination of moderate temperatures, adequate rainfall, and abundant sunlight. Properly managed, fennel can yield high-quality seeds and bulbs, contributing to its economic and culinary value.

Soil Requirements

Fennel grows best in well-drained sandy loam or loamy soils that are fertile and rich in organic matter. The ideal soil pH for fennel cultivation ranges from 6.5 to 7.5. These soil conditions provide the necessary nutrients and structure for optimal growth and development of the plants.

Well-drained soils are essential for fennel cultivation, as the plant is susceptible to root diseases in waterlogged conditions. Sandy loam soils, characterized by their good drainage and aeration, are particularly suitable for fennel. The addition of organic matter, such as compost or well-rotted manure, improves soil fertility and moisture retention, promoting healthy plant growth. In areas with heavy clay soils, proper soil management practices, including deep ploughing and the creation of raised beds, are necessary to improve drainage and prevent waterlogging. Ensuring that the soil is well-drained and fertile is crucial for maximizing fennel yield and quality. Regular soil testing and amendments can help maintain the ideal soil conditions for fennel cultivation. The application of organic and inorganic fertilizers enhances soil fertility, supporting vigorous growth and high seed production.

Land Preparation and Planting

Proper land preparation is essential for successful fennel cultivation. The soil should be ploughed deeply to create a fine tilth, which provides a loose and friable texture conducive to root development. Incorporating organic manure during land preparation enhances soil fertility and supports robust plant growth. The optimal time for sowing fennel is from October to November. The seeds are sown at a rate of 8-10 kg per hectare, either by broadcasting or drilling in rows. The recommended spacing is 45-60 cm between rows and 20-30 cm within rows, allowing adequate space for plant growth and air circulation. Thinning of seedlings is necessary to maintain proper spacing and reduce competition for nutrients and water.

The depth of planting should be about 1.5-2.0 cm, ensuring that the seeds are adequately covered with soil. After sowing, it is essential to provide light irrigation to promote seed germination and early seedling establishment. Maintaining soil moisture during the initial stages is crucial for uniform germination and healthy plant growth.

Manuring and Fertilization

Fennel requires a balanced supply of nutrients for optimal growth and yield. The application of farmyard manure at a rate of 15-20 tonnes per hectare during land preparation significantly improves soil fertility and structure. In addition to organic manure, a balanced fertilization regimen is essential:

» **Nitrogen (N):** 50 kg per hectare.

» **Phosphorus (P_2O_5):** 25 kg per hectare.

» **Potassium (K_2O):** 25 kg per hectare.

Half of the nitrogen and the full dose of phosphorus and potassium should be applied as a basal dose during sowing. The remaining nitrogen is applied as a top dressing after 30 days of sowing, supporting continuous growth and development. This staged application ensures that the plants receive the necessary nutrients during critical growth phases. Regular soil testing and monitoring of plant health can help in adjusting the fertilization regimen as needed. The application of micronutrients such as zinc and boron, through foliar sprays, can further enhance plant growth and yield. Proper nutrient management is essential to maximize productivity and produce high-quality fennel seeds.

Irrigation

Fennel requires light but regular irrigation to maintain adequate soil moisture for optimal growth. The first irrigation should be applied immediately after sowing to promote seed germination and seedling establishment. Subsequent irrigations are provided at critical stages, including during vegetative growth, flowering, and seed formation. It is crucial to avoid waterlogging, as excessive moisture can lead to root diseases and poor seed quality. Proper drainage systems should be in place to prevent water accumulation in the fields. In areas with heavy rainfall or poorly drained soils, raised beds or ridges can help improve drainage and prevent waterlogging. In addition to regular irrigation, mulching the beds with organic materials such as straw or dried leaves helps conserve soil moisture, reduce evaporation, suppress weeds, and maintain a stable soil temperature. These benefits contribute to healthy plant growth and higher yields.

Intercultural Operations

Intercultural operations such as weeding and thinning are crucial in fennel cultivation. The first weeding is usually done 4-5 weeks after sowing, followed by additional weedings at 3-4 week intervals. Weeds compete with fennel plants for nutrients, water, and light, which can significantly reduce crop yield and quality. Regular weeding helps maintain a clean and healthy growing environment.

Thinning of seedlings is necessary to maintain proper spacing and reduce competition for nutrients and water. The recommended spacing is 20-30 cm between plants, allowing adequate space for growth and air circulation. Proper thinning ensures uniform plant development and maximizes yield potential. Other intercultural operations include removing diseased or damaged plants, thinning out crowded plants, and ensuring proper ventilation and light penetration. These practices help in maintaining a healthy crop and preventing the spread of pests and diseases.

Harvesting and Postharvest Management

Fennel is typically ready for harvest 150-180 days after sowing when the seeds turn light brown and are fully matured. The plants are usually harvested by hand, cutting the entire plant and allowing it to dry in the field for a few days. Once the plants are dry, they are threshed to separate the seeds from the plant material.

Proper drying of the seeds is essential to reduce moisture content and prevent spoilage. The seeds should be spread out in a thin layer and dried in the sun for 4-5 days. After drying, the seeds are cleaned to remove any debris and impurities, ensuring high-quality fennel seeds. The cleaned seeds should be stored in cool, dry, and airtight conditions to maintain their flavor, aroma, and shelf life. Proper storage is crucial to prevent the seeds from absorbing moisture and becoming susceptible to mold and pests. Using moisture-proof containers and storing in a low-humidity environment helps preserve the quality of the fennel seeds.

Uses of Fennel

Culinary Uses

Fennel is an essential spice in many global cuisines, known for its sweet, anise-like flavor and distinctive aroma. Both the seeds and the bulb are used in cooking to impart a unique taste to a variety of dishes. Key culinary uses include:

1. **Flavoring Agent:** Fennel seeds are used to enhance the flavor of breads, cakes, and cookies. They are often dry roasted or fried in oil to release their aromatic oils.

2. **Vegetable:** Fennel bulbs are a popular vegetable in Mediterranean cuisine, used in salads, stews, and roasted dishes.
3. **Spice Blends:** Ground fennel is a vital component of spice blends and rubs, used to season meats, fish, and vegetables.
4. **Pickling:** Fennel seeds are used in pickling recipes, adding a distinctive flavor to preserved vegetables.
5. **Beverages:** Fennel seeds are used in herbal teas and digestive drinks, providing a refreshing and healthful beverage.

Medicinal Uses

Fennel has been valued for its medicinal properties for centuries, particularly in traditional medicine systems like Ayurveda and Traditional Chinese Medicine. It is known for its digestive, carminative, and anti-inflammatory properties. Key medicinal uses include:

1. **Digestive Aid:** Fennel seeds are widely used to relieve indigestion, bloating, and gas. They stimulate the secretion of digestive enzymes and help in the assimilation of nutrients.
2. **Anti-inflammatory:** The active compounds in fennel have anti-inflammatory effects that can help reduce pain and inflammation in conditions like arthritis.
3. **Antioxidant Properties:** Fennel seeds are rich in antioxidants, which help protect the body from oxidative stress and reduce the risk of chronic diseases.
4. **Antimicrobial:** Fennel has natural antimicrobial properties, making it useful in treating infections and promoting overall health.
5. **Respiratory Health:** Fennel is used to alleviate symptoms of respiratory conditions such as asthma and bronchitis by acting as a bronchodilator and expectorant.

Cosmetic Uses

Fennel is also utilized in the cosmetic industry for its beneficial effects on the skin and hair:

1. **Skin Care:** Fennel oil is used in skincare products for its antibacterial and anti-inflammatory properties. It helps treat acne, boils, and skin infections.
2. **Anti-Aging:** The antioxidants in fennel help fight free radicals, reducing signs of aging such as wrinkles and fine lines.
3. **Hair Care:** Fennel oil is used in hair care formulations to strengthen hair, prevent hair loss, and treat dandruff and scalp infections.

Industrial Uses

Beyond culinary and medicinal applications, fennel finds its way into various industrial uses:

1. **Flavoring Agents:** Fennel extracts and essential oils are used in the food industry to flavor sauces, marinades, and processed foods.
2. **Fragrance Industry:** Fennel oil is used in perfumes and aromatic products for its sweet, spicy scent.
3. **Natural Preservative:** Due to its antimicrobial properties, fennel is used as a natural preservative in food products to extend shelf life.

Cultural and Ritual Uses

Fennel has cultural significance in many regions, where it is used in traditional rituals and ceremonies:

1. **Religious Offerings:** In some cultures, fennel is used in religious rituals and offerings due to its sacred status.
2. **Traditional Remedies:** Folk medicine practices often include fennel as a remedy for various ailments, leveraging its therapeutic properties passed down through generations.

Varieties and Their Characteristics

Variety	Developing Institution	Special Features
RF-101	ICAR-NRCSS, Ajmer	High yield, good seed quality, suitable for Rajasthan.
Azad Saunf	CSAUA&T, Kanpur	High yield, early maturing, suitable for Uttar Pradesh.
HF-33	Haryana Agricultural University, Hisar	High yield, tolerant to pests and diseases.
Gujarat Fennel-1	Gujarat Agricultural University	High yield, bold seeds, suitable for Gujarat.
Rajendra Surbhi	RAU, Bihar	High yield, good seed quality, suitable for Bihar.
CO-1	Tamil Nadu Agricultural University	Early maturing, high yield, good seed quality.
Punjab Fennel-1	Punjab Agricultural University	High yield, resistant to wilt and blight.

Area and Production of Fennel

State	Area (hectares)	Production (tonnes)	Major Varieties Grown
Rajasthan	40,000	35,000	RF-101, Gujarat Fennel-1
Gujarat	30,000	25,000	Gujarat Fennel-1
Uttar Pradesh	20,000	18,000	Azad Saunf
Punjab	10,000	9,000	Punjab Fennel-1
Bihar	5,000	4,500	Rajendra Surbhi
Haryana	5,000	4,500	HF-33
Tamil Nadu	3,000	2,500	CO-1
Madhya Pradesh	2,000	1,800	Local varieties

6. Black Pepper (*Piper nigrum* L.)

Introduction

Black pepper, often referred to as the "King of Spices," is one of the most significant and widely used spices globally. Derived from the dried berries of the pepper plant, *Piper nigrum*, it belongs to the Piperaceae family. Black pepper is renowned for its pungent flavor, aromatic profile, and various medicinal properties. It plays a crucial role in culinary traditions worldwide and holds significant economic importance in the spice trade. The spice is utilized in whole, cracked, or ground forms to season and enhance the flavor of a wide array of dishes, making it indispensable in both home kitchens and professional culinary settings.

Origin and Distribution

Black pepper is indigenous to the Western Ghats of India, a region known for its rich biodiversity and favorable climate for pepper cultivation. The spice has been cultivated and traded for thousands of years, making it one of the oldest known spices. Historical records indicate that black pepper was highly valued in ancient Greek and Roman societies, often used as currency and a valuable trade commodity.

Today, black pepper is cultivated extensively in tropical regions around the world. Major producing countries include India, Vietnam, Indonesia, Brazil, and Sri Lanka. Each of these countries has developed unique cultivation practices suited to their specific climates and soil conditions. For instance, Vietnam has become the largest producer and exporter of black pepper, contributing significantly to the global supply. The widespread cultivation and production of black pepper underscore its universal appeal and economic importance.

Botany

Black pepper is a perennial climbing vine that exhibits distinct botanical characteristics:

» **Family:** Piperaceae

- **Plant Type:** Perennial climbing vine, which can grow to significant heights when supported.
- **Leaves:** Simple, alternate, dark green, and ovate to heart-shaped. The leaves are glossy and have a characteristic aroma when crushed.
- **Flowers:** Small, white to greenish, arranged in spikes known as catkins. These flowers are typically inconspicuous but play a crucial role in fruit production.
- **Fruits:** Drupes, initially green, turning red upon ripening, and black when dried. The drupes, also known as peppercorns, contain a single seed and are the source of black pepper.

The pepper plant thrives in warm, humid environments and requires support to climb and grow effectively. The plant's climbing habit allows it to utilize vertical space efficiently, making it suitable for cultivation in plantation systems where space optimization is crucial. The fruits undergo significant color changes as they mature, which is a key indicator for determining the optimal harvest time.

Climatic Requirements

Black pepper thrives in tropical climates characterized by high humidity and consistent rainfall. The ideal temperature range for black pepper cultivation is between 25°C and 30°C. These temperatures are conducive to vigorous vegetative growth and optimal fruit production. The plant requires an annual rainfall of 2000-3000 mm, which should be well-distributed throughout the year to maintain adequate soil moisture levels. Black pepper is typically grown at altitudes ranging from 150 to 1500 meters above sea level. These altitudes provide the necessary climatic conditions, including temperature and humidity, that support healthy growth and development.

Regions with tropical climates, such as parts of India, Vietnam, Indonesia, Brazil, and Sri Lanka, provide ideal growing conditions for black pepper. High humidity levels and frequent rainfall are essential for the plant's growth, as they help maintain the moisture levels required for vigorous vegetative and reproductive growth. Ensuring these climatic conditions can lead to higher yields and better-quality peppercorns.

Soil Requirements

Black pepper grows best in well-drained loamy soils that are rich in organic matter. The ideal soil pH for black pepper cultivation ranges from 4.5 to 6.5. These soil conditions provide the necessary nutrients and structure for optimal growth and development of the plants.

Well-drained soils are essential for black pepper cultivation, as the plant is susceptible to root diseases in waterlogged conditions. Loamy soils, characterized by their good drainage and aeration, are particularly suitable for black pepper. The addition of organic matter, such as compost or well-rotted manure, improves soil fertility and moisture retention, promoting healthy plant growth. In areas with heavy clay soils, proper soil management practices, including deep ploughing and the creation of raised beds, are necessary to improve drainage and prevent waterlogging. Ensuring that the soil is well-drained and fertile is crucial for maximizing black pepper yield and quality.

Regular soil testing and amendments can help maintain the ideal soil conditions for black pepper cultivation. The application of organic and inorganic fertilizers enhances soil fertility, supporting vigorous growth and high fruit production.

Land Preparation and Planting

Proper land preparation is essential for successful black pepper cultivation. The land should be cleared of any debris and vegetation, and the soil should be prepared to a fine tilth by deep ploughing. Adding organic matter, such as compost or farmyard manure, improves soil fertility and structure, providing a conducive environment for root development.

The optimal time for planting black pepper is from June to August, coinciding with the monsoon season. This timing ensures that the young plants receive adequate moisture for establishment. Black pepper is typically propagated through stem cuttings taken from healthy, high-yielding vines. The cuttings should be 30-45 cm long, with at least 3-4 nodes, and treated with a rooting hormone to promote root development.

The cuttings are planted in pits or trenches filled with a mixture of soil and organic matter. The recommended spacing is 2.5-3.0 meters between plants to allow sufficient space for growth and air circulation. Supporting structures, such as trees, poles, or trellises, are essential for training the vines as they grow.

Manuring and Fertilization

Black pepper requires a balanced supply of nutrients for optimal growth and yield. The application of farmyard manure at a rate of 10-15 kg per plant annually during land preparation significantly improves soil fertility and structure. In addition to organic manure, a balanced fertilization regimen is essential:

» **Nitrogen (N):** 100 g per plant annually.
» **Phosphorus (P_2O_5):** 40 g per plant annually.
» **Potassium (K_2O):** 140 g per plant annually.

The application should be split into two doses, one at the onset of the monsoon season and the other during the post-monsoon period. This staged application ensures that the plants receive the necessary nutrients during critical growth phases.

Regular soil testing and monitoring of plant health can help in adjusting the fertilization regimen as needed. The application of micronutrients such as zinc and boron, through foliar sprays, can further enhance plant growth and yield. Proper nutrient management is essential to maximize productivity and produce high-quality black pepper.

Irrigation

Black pepper requires regular irrigation to maintain adequate soil moisture, especially during dry spells. The first irrigation should be applied immediately after planting to promote root establishment. Subsequent irrigations are provided at regular intervals, particularly during critical stages such as flowering and fruit setting. Drip irrigation is the preferred method for black pepper cultivation, as it ensures efficient water use and maintains optimal soil moisture levels. Drip irrigation also helps prevent waterlogging, which can lead to root diseases and poor fruit quality. Proper drainage systems should be in place to prevent water accumulation in the fields. In addition to regular irrigation, mulching the plantation with organic materials such as straw or dried leaves helps conserve soil moisture, reduce evaporation, suppress weeds, and maintain a stable soil temperature. These benefits contribute to healthy plant growth and higher yields.

Intercultural Operations

Intercultural operations such as weeding and mulching are crucial in black pepper cultivation. Regular weeding helps maintain a clean plantation, reducing competition for nutrients and water. Weeding should be done carefully to avoid damaging the shallow root system of the pepper plants.

Mulching with organic materials such as straw or dried leaves helps conserve soil moisture, improve soil fertility, and suppress weed growth. Mulching also helps maintain a stable soil temperature, promoting healthy root development. Training and pruning are essential practices in black pepper cultivation. Training involves guiding the vines to grow on supports such as trees, poles, or trellises. This practice helps maximize sunlight exposure and air circulation, which are crucial for healthy growth and high yields. Pruning involves removing dead or diseased branches and encouraging the growth of healthy shoots. Pruning should be done regularly to maintain the structure and vigor of the vines. Proper training and pruning practices contribute to the overall health and productivity of the black pepper plants.

Harvesting and Postharvest Management

Black pepper is typically ready for harvest 6-8 months after flowering, when the berries turn red and are fully matured. The berries are harvested by hand, picking the entire spike to ensure uniform ripening. Harvesting is usually done in the morning when the berries are firm and less prone to damage.

After harvesting, the berries are sun-dried for 7-10 days until they turn black and wrinkled. Proper drying is essential to reduce moisture content and prevent spoilage. The dried berries are then threshed to separate the seeds from the spikes. The cleaned seeds should be stored in cool, dry, and airtight conditions to maintain their flavor, aroma, and shelf life. Proper storage is crucial to prevent the seeds from absorbing moisture and becoming susceptible to mold and pests. Using moisture-proof containers and storing in a low-humidity environment helps preserve the quality of the black pepper seeds.

Uses of Black Pepper

Culinary Uses

Black pepper is a cornerstone spice in global cuisines, valued for its pungent flavor and aromatic properties. Key culinary uses include:

1. **Seasoning:** Black pepper is used to season a wide variety of dishes, including meats, vegetables, soups, and stews. It is often added at the end of cooking to preserve its flavor.
2. **Spice Blends:** Ground black pepper is a vital component of many spice blends, including curry powders, garam masala, and seasoning mixes.
3. **Marinades:** Black pepper is used in marinades for meats and seafood, enhancing the flavor and adding a subtle heat.
4. **Condiments:** Cracked or ground black pepper is a common condiment at dining tables, used to enhance the flavor of food.

Medicinal Uses

Black pepper has been valued for its medicinal properties for centuries. It is known for its digestive, carminative, and anti-inflammatory properties. Key medicinal uses include:

1. **Digestive Aid:** Black pepper stimulates the secretion of digestive enzymes, helping to relieve indigestion, bloating, and gas.
2. **Anti-inflammatory:** The active compound piperine has anti-inflammatory effects, reducing pain and inflammation in conditions like arthritis.
3. **Antioxidant Properties:** Black pepper is rich in antioxidants, which help protect the body from oxidative stress and reduce the risk of chronic diseases.
4. **Antimicrobial:** Black pepper has natural antimicrobial properties, making it useful in treating infections and promoting overall health.
5. **Respiratory Health:** Black pepper is used to alleviate symptoms of respiratory conditions such as asthma and bronchitis by acting as a bronchodilator and expectorant.

Cosmetic Uses

Black pepper is also utilized in the cosmetic industry for its beneficial effects on the skin and hair:

1. **Skin Care:** Black pepper oil is used in skincare products for its antibacterial and anti-inflammatory properties. It helps treat acne, boils, and skin infections.
2. **Anti-Aging:** The antioxidants in black pepper help fight free radicals, reducing signs of aging such as wrinkles and fine lines.

3. **Hair Care:** Black pepper oil is used in hair care formulations to strengthen hair, prevent hair loss, and treat dandruff and scalp infections.

Industrial Uses

Beyond culinary and medicinal applications, black pepper finds its way into various industrial uses:

1. **Flavoring Agents:** Black pepper extracts and essential oils are used in the food industry to flavor sauces, marinades, and processed foods.
2. **Fragrance Industry:** Black pepper oil is used in perfumes and aromatic products for its warm, spicy scent.
3. **Natural Preservative:** Due to its antimicrobial properties, black pepper is used as a natural preservative in food products to extend shelf life.

Cultural and Ritual Uses

Black pepper has cultural significance in many regions, where it is used in traditional rituals and ceremonies:

1. **Religious Offerings:** In some cultures, black pepper is used in religious rituals and offerings due to its sacred status.
2. **Traditional Remedies:** Folk medicine practices often include black pepper as a remedy for various ailments, leveraging its therapeutic properties passed down through generations.

Varieties and Their Characteristics

Variety	Developing Institution	Special Features
Panniyur-1	Pepper Research Station, Panniyur	High yield, large berries, suitable for Kerala
Kottanadan	Kerala Agricultural University	High yield, disease-resistant, bold berries
Karimunda	Kerala Agricultural University	High yield, adaptable to various conditions
Panniyur-2	Pepper Research Station, Panniyur	High yield, drought-tolerant, large berries
Sreekara	Kerala Agricultural University	High yield, disease-resistant, bold berries
Subhakara	Kerala Agricultural University	High yield, adaptable to various conditions
Panniyur-3	Pepper Research Station, Panniyur	High yield, large berries, suitable for Kerala

Area and Production of Black Pepper

State	Area (hectares)	Production (tonnes)	Major Varieties Grown
Kerala	80,000	40,000	Panniyur-1, Kottanadan
Karnataka	30,000	15,000	Karimunda, Panniyur-2
Tamil Nadu	20,000	10,000	Sreekara, Subhakara
Andhra Pradesh	10,000	5,000	Panniyur-3, Karimunda
Odisha	5,000	2,500	Local varieties
Assam	5,000	2,500	Local varieties

7. Cardamom (*Elettaria cardamomum* Maton)

Introduction

Cardamom, often referred to as the "Queen of Spices," is renowned for its unique aroma and flavor. This highly valued spice belongs to the Zingiberaceae family and is native to the evergreen forests of the Western Ghats in South India. It plays a significant role in culinary applications, traditional medicine, and as a flavoring agent in beverages and sweets. The distinctive taste and fragrant scent of cardamom make it a staple in both sweet and savory dishes, particularly in South Asian, Middle Eastern, and Scandinavian cuisines.

Beyond its culinary uses, cardamom holds an esteemed place in traditional medicine. Its seeds and oil are utilized for their therapeutic properties, including digestive, carminative, and anti-inflammatory benefits. The spice is also a key ingredient in various traditional remedies aimed at treating respiratory issues, digestive disorders, and infections.

Origin and Distribution

Cardamom is indigenous to the Western Ghats of South India, a region known for its rich biodiversity and favorable climate for spice cultivation. Historical records suggest that cardamom has been cultivated and traded since ancient times, making it one of the oldest known spices. Today, cardamom is primarily cultivated in India, Sri Lanka, Guatemala, and Tanzania. India and Guatemala are the leading producers of cardamom, contributing significantly to the global supply.

The extensive cultivation of cardamom in these regions is due to their suitable climatic conditions and established agricultural practices. In India, cardamom is predominantly grown in the states of Kerala, Karnataka, and Tamil Nadu. Similarly, Guatemala has become a major producer due to its conducive high-altitude tropical climate. The spice's global distribution highlights its adaptability and the high demand for its unique flavor and medicinal properties.

Botany

Cardamom is a perennial herbaceous plant with several distinctive botanical features:

- **Family:** Zingiberaceae
- **Plant Type:** Perennial herbaceous plant
- **Leaves:** Lanceolate, alternate, and leathery, with a smooth texture and prominent veins. The leaves are arranged in two rows, creating a lush green canopy.
- **Flowers:** Small, white to pale violet, arranged in loose spikes. Each flower consists of three petals and sepals, and the flowering spikes emerge from the base of the plant.
- **Fruits:** Capsules, triangular in cross-section, containing small, aromatic seeds. The capsules are green when immature, turning yellow or brown upon ripening.

Cardamom plants can grow up to 3-4 meters in height, forming dense clumps through rhizome propagation. The plant's structure, including its robust root system and extensive leaf coverage, supports its growth in shaded, humid environments. The aromatic seeds, housed within the fruit capsules, are the primary commercial product of the plant.

The flowers of cardamom are not only beautiful but also crucial for fruit production. Pollination is typically carried out by bees and other insects, which are attracted to the flowers' nectar. Successful pollination results in the development of the characteristic cardamom capsules.

Climatic Requirements

Cardamom thrives in tropical climates with high humidity and well-distributed rainfall. The ideal temperature range for cardamom cultivation is between 10°C and 35°C. These temperatures support vigorous vegetative growth and optimal fruit production.

Cardamom requires an annual rainfall of 1500-4000 mm, which should be well-distributed throughout the year to maintain adequate soil moisture levels. The plant prefers shaded conditions and is often grown under the canopy of forest trees or intercropped with other perennial crops. Cardamom is typically grown at altitudes ranging from 600 to 1500 meters above sea level. These altitudes provide the necessary climatic conditions, including temperature and humidity, that support healthy growth and development. Ensuring these climatic conditions can lead to higher yields and better-quality capsules.

Soil Requirements

Cardamom grows best in well-drained, loamy soils that are rich in organic matter. The ideal soil pH for cardamom cultivation ranges from 4.5 to 6.5. These soil conditions provide the necessary nutrients and structure for optimal growth and development of plants. Well-drained soils are essential for cardamom cultivation, as the plant is susceptible to root diseases in waterlogged conditions. Loamy soils, characterized by their good drainage and aeration, are particularly suitable for cardamom. The addition of organic matter, such as compost or well-rotted manure, improves soil fertility and moisture retention, promoting healthy plant growth. In areas with heavy clay soils, proper soil management practices, including deep ploughing and the creation of raised beds, are necessary to improve drainage and prevent waterlogging. Ensuring that the soil is well-drained and fertile is crucial for maximizing cardamom yield and quality. Regular soil testing and amendments can help maintain the ideal soil conditions for cardamom cultivation. The application of organic and inorganic fertilizers enhances soil fertility, supporting vigorous growth and high fruit production.

Land Preparation and Planting

Proper land preparation is essential for successful cardamom cultivation. The land should be cleared of any debris and vegetation, and the soil should be prepared to a fine tilth by deep ploughing. Adding organic matter, such as compost or farmyard manure, improves soil fertility and structure, providing a conducive environment for root development.

The optimal time for planting cardamom is from June to August, coinciding with the monsoon season. This timing ensures that the young plants receive adequate moisture for establishment. Cardamom is typically propagated through rhizome cuttings with at least one healthy sprout. The cuttings should be 15-20 cm long and treated with a fungicide to promote healthy growth. The cuttings are planted in pits or trenches filled with a mixture of soil and organic matter. The recommended spacing is 2 x 2 meters or 2 x 1.5 meters, depending on the variety and planting system. Proper spacing ensures adequate air circulation and sunlight penetration, which are essential for healthy growth.

Manuring and Fertilization

Cardamom requires a balanced supply of nutrients for optimal growth and yield. The application of farmyard manure at a rate of 10-15 kg per plant annually during land preparation significantly improves soil fertility and structure. In addition to organic manure, a balanced fertilization regimen is essential:

» **Nitrogen (N):** 75 g per plant annually.

» **Phosphorus (P_2O_5):** 75 g per plant annually.

» **Potassium (K_2O):** 150 g per plant annually.

The application should be split into three doses: one at the onset of the monsoon, one during the mid-monsoon, and one post-monsoon. This staged application ensures that the plants receive the necessary nutrients during critical growth phases.

Regular soil testing and monitoring of plant health can help in adjusting the fertilization regimen as needed. The application of micronutrients such as zinc and boron, through foliar sprays, can further enhance plant growth and yield. Proper nutrient management is essential to maximize productivity and produce high-quality cardamom capsules.

Irrigation

Cardamom requires regular irrigation to maintain adequate soil moisture, especially during dry spells. The first irrigation should be applied immediately after planting to promote root establishment. Subsequent irrigations are provided at regular intervals, particularly during critical stages such as flowering and fruit setting. Drip irrigation is the preferred method for cardamom cultivation, as it ensures efficient water use and maintains optimal soil moisture levels. Drip irrigation also helps prevent waterlogging, which can lead to root diseases and poor fruit quality. Proper drainage systems should be in place to prevent water accumulation in the fields. In addition to regular irrigation, mulching the plantation with organic materials such as straw or dried leaves helps conserve soil moisture, reduce evaporation, suppress weeds, and maintain a stable soil temperature. These benefits contribute to healthy plant growth and higher yields.

Intercultural Operations

Intercultural operations such as weeding and mulching are crucial in cardamom cultivation. Regular weeding helps maintain a clean plantation, reducing competition for nutrients and water. Weeding should be done carefully to avoid damaging the shallow root system of the cardamom plants.

Mulching with organic materials such as straw or dried leaves helps conserve soil moisture, improve soil fertility, and suppress weed growth. Mulching also helps maintain a stable soil temperature, promoting healthy root development. Training and pruning are essential practices in cardamom cultivation. Training involves guiding the plants to remove dead or diseased shoots and encourage healthy growth. Pruning is typically done after harvesting to remove old and unproductive shoots, promoting the growth of new, healthy shoots. Proper training and pruning practices contribute to the overall health and productivity of the cardamom plants. Regular monitoring and maintenance ensure that the plants remain vigorous and capable of producing high-quality capsules.

Harvesting and Postharvest Management

Cardamom is typically ready for harvest 2-3 years after planting, usually during the months of October to December. The capsules are harvested by hand, picking the mature capsules when they are full-grown but still green. Harvesting is done in the morning when the capsules are firm and less prone to damage. After harvesting, the capsules are dried to reduce moisture content and prevent spoilage. Sun-drying is a common method, where the capsules are spread out in thin layers and dried for 3-4 days. Alternatively, drying chambers can be used to ensure uniform drying and retain the green color of the capsules. Proper drying is essential to reduce moisture content and preserve the quality of the capsules. The dried capsules are then threshed to separate the seeds from the spikes. The cleaned seeds should be stored in cool, dry, and airtight conditions to maintain their flavor, aroma, and shelf life. Proper storage is crucial to prevent the seeds from absorbing moisture and becoming susceptible to mold and pests.

Uses of Cardamom

Culinary Uses

Cardamom is an essential spice in many global cuisines, known for its sweet, floral flavor and distinctive aroma. Both the seeds and the pods are used in cooking to impart a unique taste to a variety of dishes. Key culinary uses include:

1. **Flavoring Agent:** Cardamom pods and seeds are used to flavor both sweet and savory dishes, including curries, rice dishes, and desserts. It is a key ingredient in many spice blends, such as garam masala and chai tea.
2. **Baking:** Ground cardamom is used in baking to add a warm, aromatic flavor to cakes, cookies, and pastries.
3. **Beverages:** Cardamom is used to flavor a variety of beverages, including coffee, tea, and liqueurs. It is a common ingredient in spiced teas and mulled wines.
4. **Pickling:** Cardamom seeds are used in pickling recipes to add a distinctive flavor to preserved vegetables and fruits.

Medicinal Uses

Cardamom has been valued for its medicinal properties for centuries, particularly in traditional medicine systems like Ayurveda and Traditional Chinese Medicine. It is known for its digestive, carminative, and anti-inflammatory properties. Key medicinal uses include:

1. **Digestive Aid:** Cardamom seeds and oil are widely used to relieve indigestion, bloating, and gas. They stimulate the secretion of digestive enzymes and help in the assimilation of nutrients.
2. **Anti-inflammatory:** The active compounds in cardamom have anti-inflammatory effects that can help reduce pain and inflammation in conditions like arthritis.
3. **Antioxidant Properties:** Cardamom seeds are rich in antioxidants, which help protect the body from oxidative stress and reduce the risk of chronic diseases.
4. **Antimicrobial:** Cardamom has natural antimicrobial properties, making it useful in treating infections and promoting overall health.
5. **Respiratory Health:** Cardamom is used to alleviate symptoms of respiratory conditions such as asthma and bronchitis by acting as a bronchodilator and expectorant.

Cosmetic Uses

Cardamom is also utilized in the cosmetic industry for its beneficial effects on the skin and hair:

1. **Skin Care:** Cardamom oil is used in skincare products for its antibacterial and anti-inflammatory properties. It helps treat acne, boils, and skin infections.

2. **Anti-Aging:** The antioxidants in cardamom help fight free radicals, reducing signs of aging such as wrinkles and fine lines.
3. **Hair Care:** Cardamom oil is used in hair care formulations to strengthen hair, prevent hair loss, and treat dandruff and scalp infections.

Industrial Uses

Beyond culinary and medicinal applications, cardamom finds its way into various industrial uses:

1. **Flavoring Agents:** Cardamom extracts and essential oils are used in the food industry to flavor sauces, marinades, and processed foods.
2. **Fragrance Industry:** Cardamom oil is used in perfumes and aromatic products for its sweet, spicy scent.
3. **Natural Preservative:** Due to its antimicrobial properties, cardamom is used as a natural preservative in food products to extend shelf life.

Cultural and Ritual Uses

Cardamom has cultural significance in many regions, where it is used in traditional rituals and ceremonies:

1. **Religious Offerings:** In some cultures, cardamom is used in religious rituals and offerings due to its sacred status.
2. **Traditional Remedies:** Folk medicine practices often include cardamom as a remedy for various ailments, leveraging its therapeutic properties passed down through generations.

Varieties and Their Characteristics

Variety	Developing Institution	Special Features
Malabar	ICRI (Indian Cardamom Research Institute)	High yield, adaptable to varying conditions
Mysore	ICRI	High yield, large capsules, bold seeds
Vazhukka	Kerala Agricultural University	High yield, disease-resistant, bold capsules
Njallani	Kerala Agricultural University	High yield, large capsules, suitable for Kerala
PV-1	ICRI	High yield, drought-tolerant, large capsules
Green Gold	Kerala Agricultural University	High yield, disease-resistant, bold capsules
Mudigere 1	ICRI	High yield, adaptable to varying conditions

Area and Production of Cardamom

State	Area (hectares)	Production (tonnes)	Major Varieties Grown
Kerala	40,000	20,000	Malabar, Njallani
Karnataka	20,000	10,000	Mysore, PV-1
Tamil Nadu	10,000	5,000	Vazhukka, Green Gold
Andhra Pradesh	5,000	2,500	Njallani, Mudigere 1
Odisha	2,000	1,000	Local varieties
Assam	2,000	1,000	Local varieties

References

A guide on vegetable crops (D.Veeraragavatam, M.Jawharlal and Seemanthi Ramada),Suri Associates, Coimbatore,1998.

Advances in horticulture, vol. V.(K.L.Chadha and G.Kallo), Malhotra publishing house, New Delhi, 1993.

Advances in tomato technology (Rai Kanta Matti, Humberto Gonzalez –Radriguez, N.C.Sarkar, Vidyasagar Parchuri), New Delhi Publishers,2009.

Botany of vegetable crops (S. Subramaniaan, D. Veeraraghavatham, P. Jansiram and S.Thamburaj) , KRS Sareen printers, Coimbatore.

Cucurbits (N.M. Nayar and T.A. More), IBH Publishing Co, New Delhi ,1998.

Cucurbits, botany, cultivation and utilization (T W Whitaker G.N.Davis) , Interscience, Newyork,1962.

Fertilizer management in food crops (H.L.S.Tandon),Fertilizer Development and consultation organization, New Delhi,1993.

Hand book of vegetable science (Nagendra Rai and Bhagwat Saran Asati), International Book Distributing Co, Lucknow, 2008.

Indian Horticulture Database (J.P. Negi , Mitra Lily H.K. Dabas) , National Horticulture Board , Gurgaon, 1998.

Manurial requirements of vegetable crops (Singh Kirti), ICAR, New Delhi, 1991.

Modern technology in vegetable production (P.Hazra, A. Chattopadhyay, K.Karmakar and S.Dutta), New India Publishing Agency, New Delhi, 2011

Modern technology on vegetable production (Nem pal singh, A.K. Bhardwaj , Abnish kumar and K.M.Singh) International Book Distributing Co, Lucknow, 2004.

Principals of vegetable production (S.P. Singh) Agrotech Publishing Academy, Udaipur, 2006.

Production technology of vegetable crops (S.P.Singh), Agric. Res.Comm. Center, Karnal, 1989.

Production technology of vegetable crops (K.G. Shanmugavelu), Oxford and IBH, 1993.

Production Technology of Vegetable crops (K.G. Shanmugavelu), Oxford &IBH, 1993.

Technology for vegetable production and improvement (P. Hazra and M. G. Som), Naya prakash, Kolkata. West Bengal, 1999.

Text book of vegetables, tuber crops and spices (eds. S. Thamburaj and Narendra singh), I.C.A.R., New Delhi, 2001.

The Physiology of vegetable crops (ed. H.C. Wein), CAB international, UK, USA, 1997.

The Physiology of vegetable crops (ed. H.C. Wien), CAB international , UK, USA, 1997.

Varieties of vegetable crops in India ((P. Hazra and M.K.Banerjee), Kalyani publishers, Ludhiana, 2005.

Vegetable crops (H.C. Thomson and W.C. Kelly), Tata McGraw Hill, 1979.

Vegetable crops (T. R. Gopalakrishnan), New India Publishing Agency, New Delhi,2007.

Vegetable crops in India (eds. T.K. Bose and M.G. Som), Naya Prakash, 1986.

Vegetable production in India (D. V. S. Chauhan), Ramprasad and sons, Agra, Bhopal,1981.

Vegetable Science (P. Hazra and M.G. Som), Kalyani publishers, Ludhiana, 2006.

Vegetable Science and technology in India (V. Swarup) , Kalyani publishers, Ludhiana, 2006.

Vegetable crop production (K. V. Kamath), Oxford Book Company , Jaipur, 2007.

Vegetables (B. Choudhary), National book trust, New Delhi, 1996.

Vegetables for the tropical regions (Nath Prem Velayudhan Sundari and D.P. Singh), ICAR, New Delhi, 1994

For Product Safety Concerns and Information please contact our
EU representative GPSR@taylorandfrancis.com Taylor & Francis
Verlag GmbH, Kaufingerstraße 24, 80331 München, Germany